THE CONCEPT OF MORAL CONSENSUS

Philosophy and Medicine

VOLUME 46

The titles published in this series are listed at the end of this volume.

THE CONCEPT OF MORAL CONSENSUS

The Case of Technological Interventions in Human Reproduction

Edited by

KURT BAYERTZ

Department of Philosophy, University of Münster, Germany

SPRINGER-SCIENCE+BUSINESS MEDIA, B.V.

Library of Congress Cataloging-in-Publication Data

The Concept of moral consensus : the case of technological
interventions into human reproduction / edited by Kurt Bayertz.
p. cm. -- (Philosophy and medicine ; v. 46)
Includes bibliographical references and index.
ISBN 978-94-010-4371-7 ISBN 978-94-011-0860-7 (eBook)
DOI 10.1007/978-94-011-0860-7
1. Medical ethics--Congresses. 2. Ethics committees--Decision
making--Congresses. 3. Consensus (Social sciences)--Congresses.
4. Human reproductive technology--Moral and ethical aspects-
-Congresses. I. Bayertz, Kurt. II. Series.
R725.3.C66 1994
176--dc20 93-40700

ISBN 978-94-010-4371-7

printed on acid-free paper

TABLE OF CONTENTS

PREFACE

Books do not come about by accident. This is especially the case when a volume grows out of a conference for which the participants wrote the original contributions in different languages. This volume descends from a conference held at the *Zentrum für interdisziplinäre Forschung*, University of Bielefeld, Germany, October 4 through 6, 1990, under the title "Technische Eingriffe in die menschliche Reproduktion: Perspektiven eines moralischen Konsenses". Many with great generosity helped to ensure that the conference was a success and that the papers presented grew into a book. We want in particular to acknowledge our deep gratitude to the *Zentrum für interdisziplinäre Forschung* for sponsoring this important conference, and to its director, Peter Weingart, for his important guidance and support. Our thanks are also due to all of the staff of the Zentrum. It is they who made the conference successful. We are also grateful to Prof. Hilmar Stolte, head of the *Institut für System- und Technologieanalysen* in Bad Oeynhausen, Germany, for making available additional financial support for the conference. Our thanks are also owed to the participants who inspired us to transform a collection of papers into a completed volume. The general transformation of the original papers required translation. Here we must acknowledge the labors of Sarah L. Kirkby, who rendered many parts of the volume into English. Finally, we want to recognize the invaluable support given by the ecumenical teamwork of Kurt W. Schmidt and Kevin Wm. Wildes, S.J., who managed to compose a respectable text from the various manuscripts received. Many others helped in many ways and we express our indebtedness to them as well.

KURT BAYERTZ
H. TRISTRAM ENGELHARDT, JR.

KURT BAYERTZ

INTRODUCTION: MORAL CONSENSUS AS A SOCIAL AND PHILOSOPHICAL PROBLEM*

I. A MORAL CRISIS?

1. A demand for moral consensus arises due to the latter's absence. Whichever problem we choose to examine in connection with the practical aspects of moral life, even if it is one of relatively minor significance, for each opinion held there will be another to counter it, and for each approach to problem solving an alternative will be suggested. It hardly needs to be explained why philosophical reflection on morality does not take place more harmoniously: the discipline of contemporary ethics is characterized by a vast diversity – or, if preferred, a chaos – of heterogeneous theories and concurring approaches. This absence of moral consensus is usually considered a state of crisis. A decade ago, Alasdair MacIntyre diagnosed grave "disorders of moral thought and practice" in modern society:

> The most striking feature of contemporary moral utterance is that so much of it is used to express disagreements; and the most striking feature of the debates in which these disagreements are expressed is their interminable character ... I do not mean by this just that such debates go on and on – although they do – but also that they apparently can find no terminus. There seems to be no rational way of securing moral agreement in our culture ([11], p. 6).

MacIntyre describes an important characteristic of our moral life so obviously fitting that nobody could seriously contradict it. Other authors have also written about a "moral crisis" of the present day, and connected it with the decay "of ethical and metaphysical consensus" in the modern age ([6], p. 3).

Kurt Bayertz (ed.), The Concept of Moral Consensus, 1–15.

2. The problems surrounding the legitimacy of technical interventions in human reproduction may be regarded as a paradigm for this predominance of moral dissent in modern societies. The heart of modern reproductive technology, *in vitro fertilization*, has been the subject of controversial discussion for one and a half decades. While IVF is meanwhile being practiced worldwide and has already led to the birth of tens of thousands of "test tube babies", the Roman Catholic Church, for example, still continues its strict condemnation of this manner of human procreation [4], and in a public poll in the Swiss canton of Basel-Stadt, in March 1991, an absolute majority (62%) voted *in favor* of a law virtually prohibiting all the possibilities offered by modern reproductive medicine. Even within the medical profession there is little agreement concerning most ethically relevant questions. During an empirical study of the attitudes of human geneticists (from 19 different countries) toward various ethical problems within their field and medical practice (exemplified by 14 clinical case studies), D.C. Wertz and J.C. Fletcher came to the following conclusion:

> We did not find the degree of international consensus that we originally anticipated. There was more variation than consensus about the 14 clinical cases ([22], p. 77).

Even therapeutic options where there can be no doubt as to their net medical benefit – as for example somatic gene therapy – are confronted with discomfort from the public and contradiction from the specialists. As emphasized in an OTA background paper on gene therapy, it cannot be assumed that sooner or later there will be an end to this state of moral disagreement:

> There is little reason to believe that differences in opinion about the appropriateness of human gene therapy will resolve spontaneously, or even after extensive public discussion. With a hint of resignation: "public policy decisions will typically be made without consensus" ([20], p. 30).

In social and political practice we are not usually content with a position as varied as this. Efforts are being made everywhere to overcome the lack of agreement, at least in some questions. Here too, the moral legitimacy of technical intervention in human reproduction may be regarded as paradigmatic. Since the first "test tube baby" was born on July 25, 1978, attempts have been made throughout the world to reach, with the help of commissions, consensus about the moral and legal questions involved in gene and reproductive technologies. The first such undertaking was reflected in the work of the Warnock Committee in Great Britain, and many others have followed. Even international bodies, such as the European Parliament, have expressed opinions or appointed appropriate committees to reflect on these problems. Sometimes these efforts are directed to future legislation or to advise state bodies. In other cases they seek to improve professional practice (e.g., the *Consensus Development Program of the NIH*, cf. [16]). In yet other cases, social groups seek to establish internal consensus. In short, the predominance of dissent is perceived as a problem that has to be overcome.

3. Perception of a moral problem – and especially a moral crisis – is often cause for philosophical reflection. Astonishingly, concerning the concept of consensus this has rarely been the case. This book attempts, at least partly, to bridge this gap. It focuses on the moral problems surrounding technical intervention in human reproduction and the difficulties of achieving social consensus regarding an appropriate solution to these problems. At the same time, the contributions in this book raise the principal problems surrounding the formation of consensus in modern societies. They take technical intervention in human reproduction as an exemplary case, on the basis of which various, far reaching, fundamental ethical questions may be discussed.

The first problematic dimension shared by a number of authors is of an *explicative* nature. It refers to the search for a more precise definition of consensus. With an initial definition of 'consensus' as "interindividual agreement", various questions arise. For example, is "agreement" a result or a process? Consider three important facets of the concept of consensus:

(1) Who agrees?
(2) What is agreed upon?
(3) How is agreement reached?

The third question addresses another complex problem that is of an *evaluative* nature, and refers to the moral status of consensus. Considering the judgment, usually taken for granted, that dissent is bad and consensus is good, to ask what is special and valuable about intersubjective agreement may seem strange. Yet is it not possible that the search for agreement is no more than a psychologically relevant or politically beneficial longing for harmony? However, something which is psychologically relevant or politically beneficial is not necessarily ethically significant. Thus we have to examine the moral authority behind consensus.

II. SUBJECT AND OBJECT

4. Consider the first of the three questions posed: "Who agrees?" It is aimed at the *subject* of the consensus. It is important to distinguish here between small groups or coherent communities and large societies, states, or nations. It should be emphasized that this distinction is not primarily founded on the number of individuals involved; it has a much more structural nature. In traditional "face-to-face" communities, intersubjective agreement is guaranteed from the outset by the structure of the social relationships. The citizens know each other personally;[1] their circumstances are similar; social control is a very tightly meshed net. The individuals are raised within a powerful tradition from which they can hardly escape; often there exists neither the chance to stray from established life styles and "traditional" thinking, nor any apparent motivation to doubt the legitimacy of these lifestyles. Consensus is thus an essential feature of the "superstructure" of these communities. In contrast, modern societies seem to be collections of independent individuals and groups who share little in common. The infrastructure within such societies is brought about not by personal communication but by anonymous mechanisms and institutions. Tradition does not have a powerful influence on the lives and thoughts of individuals, and social control is less strict within most areas. This type of society can be divided into many subsocieties and subcultures, each with its own system of values. Thus they are "pluralistic": within them there exist many forms of life, life styles, and visions of "the good life". The members of these societies often encounter each other as "moral strangers" [7]. Thus the formation of moral consensus tends to become problematic in large societies.

Of course, the heterogeneity of modern societies should not be exaggerated. Even though there are few commonly shared convictions within them, there are many shared values among the often various subcultures and communities. That is, the various social groups are connected by *moral family resemblances*. Thus the search for consensus within such societies is by no means condemned to fail from the start: alongside dissent we can find consensus as well. This will not, however, reflect universal agreement; the idea of a consensus which encompasses all questions and problems, and which includes all individuals, must be abandoned. Typical for large, modern societies is the achievement of *particular consensus*: consensus about a particular question among a limited number of individuals. There is little point trying to hide the fact that this implies a significant qualification: a reduction of "consensus" to *majority*.[2] Large modern societies can, of course, "imitate" the processes of consensus formation typical of small "face-to-face" communities. In order to do this, commissions are established to give voice to a particular complex set of questions and to solve (consensually) the normative problems that arise from it.

5. The second question (What is agreed upon?) refers to the *object* of consensus. The likelihood of intersubjective agreement depends on *what* is to be agreed upon. To put it "strategically": how shall we structure moral debates so that consensus is a possible outcome? Two perceptions play an important role: for the first of these, the moral pluralism that exists within modern societies is a fact that must be accepted without regret. Moral pluralism will not overcome the existing heterogeneity and facilitate a general consensus, but aims to find a neutral position from which it is possible to live *with and within* this heterogeneity. Thus the idea of consensus about substantial norms and values must be abandoned from the beginning. Insofar as it is possible to reach consensus in the circumstances posed by the reality of existing pluralism, this can only refer to the *procedure* of approaching these manifold norms and values. Consensus is possible only in matters of procedure. This perception is discussed in great detail by several contributors (cf. [2]; [7]; [10]; [19]).

A second perception assumes that the continuance of moral pluralism and the predominance of normative dissent in modern society are the results of faulty development. According to Albert R. Jonsen and Stephen Toulmin, the specific type of modern thinking and the moral paradigm based on it are responsible for the absence of moral consensus today. They are convinced that the abstract universalism of modern ethics is the chief reason why it is impossible to abolish dissent and to formulate a judgment agreed on by all when evaluating particular concrete cases. Characteristic of this universalism is the way in which it tries to find support in universally applicable rules under which each individual case may be subsumed. Jonsen and Toulmin view the unfruitful and insoluble "battles of principle", which we are able to observe in most modern moral debates, as an unavoidable consequence of this manner of thinking and arguing. One such battle of principle occurs in the debate on abortion:

In former times there were always those who could discuss the morality of abortion temperately and with discrimination: acknowledging that here, as in other agonizing human situations, conflicting considerations are involved and that a just, if sometimes painful, balance has to be struck between different rights and claims, interests and responsibilities ... Despite this temperate and discriminating tradition, the public rhetoric of the abortion controversy has increasingly come, in recent years, to turn on "matters of principle". The more this has happened, the less temperate, less discriminating, and above all less resoluble the debate has been ([9], p. 4).

According to Jonsen and Toulmin, the problem of abortion can only be resolved if it is discussed not in *termini* of irreconcilable rights – the "right of the woman to self-determination" versus the "right of the embryo to live" – but with reference to concrete persons in concrete circumstances: should or may *this* woman in *these* circumstances undergo a pregnancy termination? However, the objects for moral reflection are then no longer universal principles or general rights, but concrete cases. The empirical foundation for Jonsen and Toulmin's view was their participation on a commission established by the U.S. Congress in 1974 in order to give voice to the ethical problems surrounding the protection of human subjects of biomedical and behavioral research. Widely diverse opinions among the members of this ideologically, ethnically, and professionally heterogeneous commission arose when concrete cases concerning individuals or groups were being debated. On this level it was also possible to formulate and assent to various recommendations. However, consensus collapsed as soon as the individual members began to discuss the *reasons* behind their decision.

Members of the commission were largely in agreement about their specific practical recommendations; they agreed what it was they agreed about; but the one thing they could not agree on was *why* they agreed about it. So long as the debate stayed on the level of particular judgments, the eleven commissioners saw things in much the same way. The moment it soared to the level of "principles", they went their separate ways ([9], p. 18).

Regardless of one's confidence in a careful ethical analysis of individual cases, a new casuistry cannot solve the consensus problem. First, concentration on an individual case cannot guarantee consensus. Jonsen and Toulmin take no account of the fact that there certainly are situations where consensus exists regarding principles and reasons but dissent exists regarding their application in a concrete case (cf. [3], p. 175; [14], pp. 208f.). Second, the brusque contrast between general principles and reasons on the one hand, and individual cases and situations on the other, hides the fact that every *founded* opinion about an individual case implies reasons, i.e., rules or principles which can be generally understood. As R.M. Hare has remarked, "All decisions except those, if any, that are completely arbitrary are to some extent decisions of principle" ([8], p. 65). Yet more than anything, concentrating on an individual case does not solve the central problem of modern societies: how to achieve social and/or political consensus concerning appropriate ways of acting. Before the advent of *in vitro* fertilization there was no need to regulate the use of this technology: legislators did not have to decide whether they should authorize it; nor did clinicians have to decide whether they should provide it; nor did public health officials have to decide whether society should pay for it. It is obvious that regulatory matters such as these cannot be decided on the basis of casuistic considerations. It is a matter of "decisions of principle," regardless of whether these decisions occur at the level of particular institutions, within a national framework, or on an international level. If moral discourse were limited to considerations of prudence on the basis of individual cases, it would automatically detach itself from the discussion of regulations. State laws and institutional regulations would be removed from moral discussions if the latter were only competent to deal with individual cases.

It should be emphasized that the tendency toward general regulations is not just a result of more recent historical social developments which must be accepted as fact; this tendency can be justified normatively. It is the political yield of a centuries-old battle against autocratic governments, their arbitrary dominion being gradually replaced by the reliability and security of a state under the rule of law. The formulation and institutionalization of *general rights* results from a growing respect for the individual and his legitimate interests. This strengthening of individual autonomy has resulted in a significant alteration of moral discourse. Whereas traditional moral discourse largely concentrated on the concept of duty and the foundation of moral obligations, in modern ethics the concept of *moral right* has virtually replaced that of duty. This is especially noticeable in discussions on gene and reproduction technologies. Again and again attempts are made to appeal to rights which would prevent certain technical interventions in human reproduction: for example, a "right to know one's natural origins," a "right not to know", or a "right to an unmanipulated genotype". Whatever opinion one may have of such appeals, certain moral goods need *principal* protection. Thus it would not be helpful to dismiss questions of principle in moral discourse and moral debates. It would not be acceptable for public health agents to decide on a case-by-case basis whether or not to cover the costs of IVF. It would be completely unacceptable if human rights were assigned to individuals according to their unique situation.

III. MORAL AUTHORITY

6. The concept of consensus is normally used in public, political, and in philosophical discussions *contrafactually* and *postulatively*. "Consensus" is typically referred to as something to be produced or created. Behind this is the conviction that consensus is good and dissent bad. Even laments over the disorders of moral life within modern societies have their roots in this assumption. Yet it has always been characteristic of philosophers to cast doubt on what is usually taken for granted. In the case of consensus it may also be useful to question the importance of and necessity for moral agreement. Let us begin with the "disorder hypothesis." One of the reasons for the difficulty of achieving consensus within modern societies is based on the fact that modern societies no longer tend to have access to a generally accepted and binding system of norms. We live in a multicultural and pluralistic society ruled not by a monotheism but a polytheism of values. There is another no less important reason for the failure to achieve consensus mentioned neither by MacIntyre nor Jonsen and Toulmin. One common cause of moral controversy is the ambiguity of so-called *empirical facts*. Controversies of this nature can often be traced to difficulties when disclosing particular facts – independent of individual cases or matters of principle. What happened in this concrete case, and what were the circumstances? What consequences must be taken into account if this or that option is prohibited or allowed? Since these are empirical questions one might think it relatively easy to reach agreement about them; precisely this is not the case:

> In actual life ... it is precisely the facts that are the most difficult to ascertain, and it is comparatively easy to lay down hypothetical judgments, of the sort: 'if the facts are so and so, then such and such ought to be done' ([17], p. 340).

If this is correct, then reaching consensus on moral questions should neither be expected nor (viewed from the outset) judged "irrational." Rather, it has to be assumed

> that on problems of morality, as on problems of law, there is considerable room for reasonable disagreement ... It is in the nature of the subject matter that on moral matters reasonable men can, on occasion, reasonably disagree ([17], pp. 340f.).

The more fragmentary the knowledge of a particular field, the greater the room for reasonable disagreement; the newer the field in question, the more fragmentary the knowledge. It can be expected from the start that new options, such as those made possible by discoveries in science and technology, will attract significant and reasonable dissent. If the morality of a society is understood as sedimented, normatively processed experience, then it should come as no surprise that a uniformly accepted morality does not exist. The options are too new and their consequences too unclear for society and its members to have gathered sufficient information and to have given them sufficient moral reflection. Whether, for example, the ever-increasing possibilities provided by prenatal diagnosis will lead to an increase in elective abortions – including cases where the infant's affliction is minor – or whether a purported decrease in tolerance toward the disabled can be accurately determined empirically. As long as reliable answers are unavailable, there remains room for controversial estimates. Moral controversies concerning human reproductive technologies originate not only, perhaps not even primarily, from the Babylonian confusion of our moral languages lamented by MacIntyre, but from a lack of knowledge and foreseeability. The controversies are not only evidence of a moral crisis, but also the expression of an epistemological one.

7. In addition, the disorder hypothesis implies that the absence of moral consensus is merely characteristic of the present, and that former times witnessed universal consensus. Is this implication correct? Was moral consensus in former times really common and far reaching? Doubts concerning this nostalgic assumption are certainly justified. Lamenting the decay of custom and the chaos of moral convictions is certainly not a privilege of the present. In Descartes' time the criticism was voiced that each person is only prepared to recognize his own principles with respect to questions of morality, so that "in this field just as many reformers as heads could be counted" ([5], pp. 38 and 100). If the *actual behavior* of people is taken as a guide to their agreement with currently reigning morality, then the past does not fare nearly as well as the present. The abortion controversy as a paradigmatic example of *modern* dissent is illustrative inasmuch as even under strict legal prohibition the

practice of abortion has never actually ceased. Until the mid-nineteenth century even infanticide seems to have been a widely practiced method of birth control ([12], pp. 43, 68f.).[3] As far as abortion is concerned, the principal difference to former times is not the actual *behavior* of people, but different *moral discourse*. Not only are abortions carried out in secret, but the moral sanction of abortion is publicly questioned. Today abortion is viewed by many as morally acceptable; it is even said to be a moral *right*. The question of consensus and dissent thus proves to be a "second order" problem: it arises whenever the moral evaluation of an action becomes the focus of public discussion.

Moral dissent is thus an expression of the *possibility* of discussing moral questions openly. If less dissent existed concerning moral matters in the past, this may be evidence that the ability to discuss moral norms did not exist, or was very limited. In retrospect, "consensus" was often not the result of voluntary agreement, but reflected the fact that true moral choice was unavailable, and that the opportunity openly to express dissent (which may well have existed) was quite restricted. The disagreement that characterizes our moral lives today is thus not primarily due to "the language of morality passing from a state of order to a state of disorder" ([11], p. 10), but should perhaps be viewed against the background of the growth of open societies, and as the expression of increased individual autonomy.

The existence of dissent reveals the fact that moral rules are no longer taken for granted but are now the object of *reflection*. Whereas in earlier societies guidelines for what was morally right or wrong were established by religious authorities, today they have become the object of critical observation and analysis. The power of reflection removes from that which seems to be self-evident its legitimizing effect. The binding nature of morality's commandments and prohibitions is no longer based on the strength of tradition alone; it also attacks those institutionalized world views (e.g. the medieval Church) which established moral norms authoratively and which, when necessary, defended them with force. The *public sphere* which has been slowly developing within European countries since the Renaissance has created a public forum in which all matters and problems of general interest can be made thematic and debated. Safeguards like freedom of thought, opinion, speech, and freedom of the press make it difficult simply to decree "consensus" and to assert it by force. Not only is deviant behavior possible, but so too is the *freedom* to subject norms and rules to debate and to question their validity. Thus the absence of social consensus should not always be viewed a loss, but also at times an *achievement*. The absence of consensus may also be the result of individuals now free from the power of a heteronomous morality.

8. This brings us to the third question mentioned above: How is agreement reached? If consensus is neither an end in itself nor an intrinsic value, then its moral value and authority depends on the way in which it arose. There are several possibilities here. Consensus may emerge from a way of life that did not allow individuals to develop alternative perspectives. Consensus could also be the "spontaneous" result of a natural, that is unplanned and unguided process: a "complex co-evolution of different factors" ([21], p. 203). *This* kind of consensus is not meant when "consensus" formation is demanded with regard to human gene and reproductive technologies, or other moral problems. Mere *factual consensus* has no claim to moral authority. As emphasized by Moreno ([13], p. 157), agreement as such about a particular proposition cannot be the basis for normative conclusions drawn for the purpose of evaluating the same proposition. Consensus has a claim to moral authority only when it is the result of communication aimed at intersubjective understanding. In a process of intersubjective understanding, those involved are not concerned with *effecting* an agreement with their partners by employing strategic means (sanctions or gratifications), but with *convincing* them of the correctness of an empirical or normative statement using arguments, not force. The type of communication which then occurs is essentially rational: it is based on the critical weighing of reasons for and against; ideally it is not mere factual consensus but *rationally founded* consensus.

The difference between merely factual and rationally founded consensus proves to be of some philosophical significance. Since Socrates, the goal of moral philosophy has been to reflect critically on what appears self-evident, and to transform it consciously in and through critical thinking. Here it is a case of transforming merely factual consensus into rationally founded consensus. In the same way that Socratic dialogue was a most important means of critical reflection, the weighing of reasons for and against is only possible within the framework of open *discourse*. How should Socratic discourse be articulated in large societies, societies that incorporate hundreds of millions of independent individuals? Like Socratic dialogue, rational discourse is only feasible in small groups – between a limited number of individuals in direct interaction and conversation. Microinstitutions such as ethics committees and similar bodies are attractive because they present an opportunity for direct communication, an exchange of arguments, and consensus formation. Besides their pragmatic function – serving as a "vehicle for resolving ethical conflict" ([10], p. 67) – they provide the hope that a common process of communication can be set in motion, one in which rationally founded consensus can be achieved through intersubjective understanding. The exchange of arguments free from non-rational influences in microinstitutions like health care ethics committees is often only an ideal. As Tancredi suggests, it is often the very *genesis* of consensual decisions which throws a dubious light on their moral validity.

> It is the combination of factors – the potential influences on the rational evaluation of information and the unusual empowerment created by ethical decisions of committees – that the argument can be strongly posed that ethical issues in reproductive technologies should not be handled through committee decisions ([18], p. 139).

This objection should be taken seriously for it draws attention to the practical difficulties and obstacles facing rational consensus formation. The rationality of moral reflection is always endangered by disturbing influences, and can only be realized by overcoming adverse circumstances; ethics committees are no exception. It would be naive to believe that a committee could *guarantee* rationality and moral advice. There is no ideal way to achieve consensus – at least not rational consensus. To put it more generally: we have to question the notion of instantaneous and absolute rationality. Again, we are confronted with an important theoretical limitation of the *concept* of consensus. The conditions which have to be fulfilled in order to bestow social consensus with moral authority are – at least partly – themselves of a moral nature. It is imperative, for example, that individuals who negotiate to reach consensus do so honestly and without coercion. Whether or not consensus justifies particular actions is dependent on whether or not the consensus itself stands up to moral criticism. Thus, not even unanimous consensus provides the Archimedean point for the justification of moral action.

IV. SUMMARY

1. Modern societies are neither homogeneous communities nor aggregates of independent individuals. Individuals and groups have many things in common, e.g., lifestyle, *Weltanschauung*, and moral convictions. This may or may not lead to universal and comprehensive consensus, but it frequently results in a network of "moral family resemblances" among individuals and groups: a "patchwork" of local dissent and consensus.

2. Social fragmentation is not merely the expression of a Babylonian confusion of moral languages; it has various reasons and causes. First, moral dissent can be traced to empirical rather than moral confusion. Second, the question arises whether dissent should always be regarded as failure. Non-agreement is often the expression of respect. This has increased throughout history for the individual. It is now possible to express controversial views in open public discussion. In short, the existence of moral dissent may at times be regarded as an achievement.

3. Consensus is usually valued and preferred because it is psychologically comforting and politically useful. However, from a philosophical point of view it is not its benefit to individuals or groups that should be examined, but its moral authority. It is not the intersubjective agreement which is ethically relevant, but its *rational* foundation. Consensus has a claim to moral authority only when it is the result of a rational communicative process aimed at intersubjective understanding and a just balancing of interests. These communicative processes are no guarantee of success; the goal of instantaneous rationality has now to be rejected. Thus, not even social consensus is able to provide an unshakeable foundation for moral action: for each instance of consensus is itself open to moral scrutiny.

Department of Philosophy
University of Münster
Germany

NOTES

* Translations by Sarah L. Kirkby, B.A. Hons. Exon.

1 Being able to survey the *polis* played a very important role in ancient social and political philosophy. According to Plato, the State should be organized in little sections, "so that, when assemblies of each of the sections take place at the appointed times, they may provide an ample supply of things requisite, and the people may fraternize with one another at the sacrifices and gain knowledge and intimacy, since nothing is of more benefit to the State than this mutual acquaintance" (*Laws*, 738D, [15], p. 361). Aristotle also emphasizes how difficult, or even impossible it is to govern a heavily populated State. "Clearly then the best limit of the population of a state is the largest number which suffices for the purposes of life, and can be taken in at a single view" (*Politics*, 1326b, [1]).

2 Characteristic here is the operationalization of the concept of consensus dealt with by Wertz and Fletcher in their empirical investigation: "Our criteria for consensus were those frequently used in legislative processes, in the absence of an accepted scientific criterion for consensus. We used a '3/4's rule' (3/4's of the respondents in each of 3/4's of countries) to define a 'strong consensus', and a '2/3's rule' (2/3's of the respondents in each of 2/3's of countries) to define a 'moderate consensus"' ([22], p. 12).

3 Morally this does not, of course, say very much. The fact that actual human behavior may disregard a norm has no bearing on the validity of that norm. It does not even exclude the existence of a consensus about that norm: throughout history, human beings have been known to steal; yet there has been (and is) a consensus which states that stealing is morally bad. It therefore should be recorded that the moral *validity* of a norm, *consensus* regarding this validity, and factual *adherence* to the norm do not necessarily coincide. Conversely, the fact that a norm is not publicly questioned does not necessarily mean that a consensus exists regarding its validity. Deviant behavior certainly can be an indication of a lack of agreement: silent acceptance is not voluntary agreement.

BIBLIOGRAPHY

1. Aristotle: 1966 (1921), *Politics. The Works of Aristotle*, vol. X, trans. under the Editorship of W.D. Ross, Oxford University Press, London, U.K. Reprint from sheets of the First Edition 1921.
2. Bayertz, K.: 1994, 'The Concept of Moral Consensus. Philosophical Reflections' in this volume, pp. 41-57.
3. Childress, J.F.: 1994, 'Consensus in Ethics and Public Policy: The Deliberations of the U.S. Human Fetal Tissue Transplantation Research Panel', in this volume, pp. 163–187.
4. Congregation for the Doctrine of the Faith: 1987, *Instruction on the Respect for Human Life in its Origin and on the Dignity of Procreation (Donum Vitae).* Vatican City, Italy.
5. Descartes, R.: 1960, *Discours de la methode.* Bilingual edition (French/German), ed. L. Gäbe, Felix Meiner, Hamburg, Germany.

6. Engelhardt, H.T., Jr.: 1986, *The Foundations of Bioethics.* Oxford University Press, New York, NY.
7. Engelhardt, H.T., Jr.: 1994, 'Consensus: How Much Can We Hope for? A Conceptual Exploration Illustrated by Recent Debates Regarding the Use of Human Reproductive Technologies', in this volume, pp. 19–40.
8. Hare, R.M.: 1970, *The Language of Morals*, Oxford University Press, London, Oxford, New York, NY.
9. Jonsen, A.R. and Toulmin, S.: 1988, *The Abuse of Casuistry. A History of Moral Reasoning.* University of California Press, Berkeley, CA.
10. Kuhse, H.: 1994, 'New Reproductive Technologies: Ethical Conflict and the Problem of Consensus', in this volume, pp. 75–96.
11. MacIntyre, A.: 1981, *After Virtue: A Study in Moral Theory.* Duckworth, London, UK.
12. McKeown, T.: 1979, *The Role of Medicine: Dream, Mirage or Nemesis?* Basil Blackwell, Oxford, UK.
13. Moreno, J.D.: 1994, 'Consensus by Committee: Philosophical and Social Aspects of Ethics Committees', in this volume, pp. 145–162.
14. Novaes, S.: 1994, 'Beyond Consensus About Principles: Decision-Making by a Genetics Advisory Board in Reproductive Medicine', in this volume, pp. 207–221.
15. Plato: 1926, *Laws.* Plato In Twelve Volumes, vol. X, trans. R.G. Bury, Loeb Classical Library, W. Heinemann, London; Harvard University Press, Cambridge, MA.
16. Rand Corporation: 1989, *Changing Medical Practice through Technology Assessment: An Evaluation of the NIH Consensus Development Program.* Washington, D.C.
17. Singer, M.G.: 1971, *Generalization in Ethics: An Essay in the Logic of Ethics, with the Rudiments of a System of Moral Philosophy.* Atheneum, New York, NY.
18. Tancredi, L.R.: 1994, 'The Empirical Limits of Consensus: Can Theory and Practice be Reconciled?', in this volume, pp. 129–141.
19. ten Have, H.A.M.J.: 1994, 'Consensus, Pluralism and Procedural Ethics', in this volume, pp. 65–74.
20. U.S. Congress, Office of Technology Assessment: 1984, *Human Gene Therapy – A Background Paper,* Washington, D.C.
21. Weingart, P.: 1994, 'Consensus by Default: The Transition from the Social Technology of Eugenics to the "Technological-Fix" of Human Genetics', in this volume, pp. 189–206.
22. Wertz, D.C.: Fletcher, J.C. (eds.): 1989, *Ethics and Human Genetics: A Cross-Cultural Perspective,* Springer, Berlin, Heidelberg, New York.

PART ONE

PHILOSPHICAL FOUNDATIONS

H. TRISTRAM ENGELHARDT, JR.

CONSENSUS: HOW MUCH CAN WE HOPE FOR?

A Conceptual Exploration Illustrated by Recent Debates Regarding the Use of Human Reproductive Technologies

Discussions of the moral and political significance of consensus go aground on the difference between consensus in small face-to-face communities such as families, clubs, and clans on the one hand, and large-scale states on the other. Lewis and Short translate the Latin consensus as "agreement, accordance, unanimity, and concord" ([17], p. 428). The Oxford English Dictionary defines the English word consensus as "agreement in opinion" or "collective unanimous opinion of a number of persons." As we will see, the difficulty is that consensus as unanimity of opinion or agreement is possible in families, clubs, clans, and small-scale organizations. It is not possible, or it is at least highly unlikely, in large-scale states as we know them. It is possible in churches where dissent entails ipso facto excommunication. However, the life of large-scale, peaceable, democratic states is one marked by minority opinion, dissent, and lack of consensus as unanimity. Consensus in the case of large-scale, peaceable, democratic states can only mean the existence of a preponderant and overwhelming majority view in a particular matter.

Consensus, in the sense of general agreement, even if short of unanimity, about a course of action or a vision of life tends to diminish political strife and increase political cooperation. Insofar as individuals value political concord, consensus is valued in itself.[1] Insofar as political cooperation is useful, consensus also receives an instrumental value.[2] The politics of consensus recommends itself to those who celebrate the consequences of consensus in the political order. This paper will not assess the general social and political consequences of consensus. Rather, it will first show why consensus about fundamental moral issues or with respect to the fundamental character of the body politic is not morally or rationally inevitable, and indeed, highly improbable.

Kurt Bayertz (ed.), The Concept of Moral Consensus, 19–40.

Any consensus, should it be achieved, is accidental and, perhaps, fortuitous. Second, the difference between consensus and unanimity will be pressed to show that the presence of consensus cannot solve the problem of authority in the post-modern world. Though consensus is practically useful in political governance (as, for example, a reliable, fully automatic machine gun is practically useful in reducing the number of one's enemies), its moral significance remains to be judged. After all, insofar as consensus is not unanimity, it can lead to the successful tyranny of a preponderant majority over an oppressed minority. As the reader will discover, as one loses the faith that one can discover a canonical moral vision that should provide the rational basis for a ruling moral consensus, one will be moved to endorse two levels of moral discourse. On the first, one will note the existence of various and competing visions of the good life and of proper human conduct. On the second level, one will develop strategies for speaking across gulfs of moral discourse. On the first level one finds numerous and divergent moral communities, each with its own consensus. On the second level one will need to seek grounds for political authority that do not require consensus, in the sense of unanimity, regarding concrete moral issues.

This essay will draw its examples from moral discussions of reproductive technologies. Because the debate regarding the propriety of the new reproductive technologies has largely been framed in the West and in terms of Western religious and cultural values, my major accent will be on them. In particular, I will draw examples from Christianity, especially Roman Catholicism, which has articulated objections to the new reproductive technologies on the grounds that they (1) violate nature and are therefore perverse, (2) improperly separate the act of intercourse from the act of reproduction, (3) illicitly objectify the creation of human life, (4) improperly render the child to be a product rather than a gift of God, and (5) often lead to the death of embryos.

My conclusions are unlikely to be satisfactory to many. On the one hand, I will conclude that, because fundamental, normative consensus in the sense of unanimity is unattainable at the political level, the state must abandon attempts to regulate reproductive technologies other than to ensure that citizens are protected against fraud and other varieties of unconsented-to harm and coercion. This conclusion will be pleasing to some political liberals. On the other hand, I will conclude that there is in fact no way rationally to disconfirm many of the religiously and culturally-based hesitations regarding the new reproductive technologies. Indeed, specific, normative, content-full accounts of the meaning of sex and reproduction are not to be derived from an ahistorical, non-culturally-embedded sense of reason. One can discover specific, normative, content-full accounts of sex and reproduction only within particular communities or moral perspectives. It is the heterogeneity of particular communities and the inability definitively to dislodge their claims that will ensure a lack of consensus and continuing disputes. This will bring some pleasure to religious conservatives. Finally, despite the cacophony of disagreements, and in the face of a yearning for consensus, I will endeavor to show that one can justify a general, albeit limited, basis for peaceable disagreement and collaboration.

I. THE ALLURE OF CONSENSUS

More than a de facto consensus, one seeks a normative consensus. One seeks a consensus that will have moral force. The reason is rather straightforward. In moral discourse one would like to be able to show that, in at least certain areas, one is warranted to use force to protect or establish certain social structures and to achieve certain moral goods. The difficulty is that there is significant disagreement about what social structures are proper and what moral goods ought to be achieved. Even if one attained unanimity regarding an issue of morals, there would still be the question whether the view affirmed was that which must be endorsed. Even if one were to agree for purposes of argument that certain social desiderata properly have a claim on those rational individuals who wish to take the moral point of view, there would still be substantial disagreement because of different rankings or weightings given to these social desiderata. For example, if all agreed that liberty, equality, security, and prosperity are social desiderata that all should affirm, the ideal social state and the character of moral obligations will differ radically, depending on the ranking one gives to these values. It does not appear to be difficult for the moral skeptic to undermine the notion that one can establish a canonical, concrete moral account of proper social structures or proper human action.

The inevitable triumph of the moral skeptic will be addressed below in section IV. Here, in preface, it is important to understand why the advent of the moral skeptic is so disastrous for contemporary social and political assumptions. If it is the case that one cannot show why rational individuals should endorse a particular understanding of social structures or proper deportment, then the use of force to establish or protect particular social structures or to restrain particular ways of acting comes into question. Moral skepticism undermines the moral authority of social and political structures.

This circumstance underscores the close tie in Western political thought between political theory and moral theory. There has been a general attempt to justify political power by appealing to a fundamental moral account. Often, this has been undertaken either through an appeal to a Divine foundation or by a prudential, utilitarian argument to the effect that a well-ordered society is of benefit for all. The latter defense of political structure finds its recent expressions in the prisoners' dilemma interpretation of Hobbes' account of the state ([14]; [12]), as well as in accounts such as David Gauthier's ([11]). But prisoners must share in common sufficient values and understandings of the world to make a common solution of the prisoner's dilemma rationally attractive. Individuals with exotic or transcendent interests will not see the rationality of most solutions.

These difficulties to the contrary notwithstanding, the discovery or imposition of a general consensus has been considered useful and valuable (the Pax Romana being a classic example).[3] Moreover, philosophers in Greece and Rome attempted more fundamental rational justifications of the state, and of particular accounts of the state, of which Plato's Republic is an exemplar. Finally, Roman law, influenced by the Stoics, came to look at the *consensus gentium* as demonstrating that humans share certain goods and understandings in common as rational embodied creatures.[4] Christianity, which in the 4th century became the established religion of the West, developed its own appeals to consensus. In establishing doctrine, it became important to identify the *consensus fidelium*, that which had been taught by the Apostles and their immediate successors always, everywhere and by all.[5] Under the influences of modernism, the notion of consensus was transformed to identify those views embraced by the faithful as an indication of the continued revelation of the Spirit. In each case, the fact that all or nearly all agreed on certain points was seen to be a rational ground for concluding that the propositions endorsed should claim the assent of individuals. In one way or the other, appeal was made to consensus under the rubric *vox populi vox dei est*. The cardinal question is whether a moral consensus can be discovered that ought and, if one is lucky, can guide public policy bearing on the development and use of reproductive technologies.

II. HUMAN REPRODUCTIVE TECHNOLOGIES: WHY A RELIGIOUS PERSPECTIVE MAKES A DIFFERENCE

In one of his unfinished poems, "Fragment of an Agon," T.S. Eliot has Sweeney remark, "Birth, and copulation, and death. That's all the facts when you come to brass tacks" ([8], p. 80). It is not just that birth, copulation, and death are major biological facts of human life. They are in addition events around which individuals and communities build the fabric of their lives, endowing these events with cardinal significance. Or more precisely, some individuals and societies, particularly individuals who live within religious or traditional communities and the societies sustained by those communities tend to see birth, copulation, and death as major passages or events in human life.

The moral understanding of reproduction taken by the Roman Catholic Church in its condemnation of third-party-assisted reproduction illustrates such an understanding. Reproduction is seen to be a God-blessed event that appropriately occurs within marriage. The conception and birth of a child are not just physical events, or physical events with moral significance, but events with transcendent religious meaning. "Human procreation requires on the part of the spouses responsible collaboration with the fruitful love of God; the gift of human life must be actualized in marriage through the specific and exclusive acts of husband and wife, in accordance with the laws inscribed in their persons and in their union.... The child has the right to be conceived, carried in the womb, brought into the world and brought up within marriage: it is through the secure and recognized relationship to his own parents that the child can discover his own identity and achieve his own proper human development" ([4], pp. 11, 23). As a result, the Roman Catholic Church has condemned most third-party-assisted, technologically mediated reproduction.

First and foremost, the Catholic Church has condemned the use of gametes from outside of marriage as a violation of the moral integrity of marriage. The use of donor sperm and/or donor ova is regarded as an objectively grave moral evil. This interpretation requires seeing marriage as more than a voluntary union, as also a relationship willed by God and set within biological, not just contractual constraints. Even when the sperm are from the husband, the Roman Catholic Church condemns this as an improper manipulation of biological functions.[6] The Roman Catholic Church has in part opposed artificial insemination from the husband (AIH) on the grounds that masturbation needed to procure the husband's sperm is a gravely perverse act.[7] Though the Roman Catholic Church developed this position in terms of a teleological biology with Aristotelian and Stoic roots, the fundamental idea is that certain sexual acts miss the goals of a Christian life. But it is not just that masturbation as a means for procuring semen is considered perverse. In addition, the new technologies such as in vitro fertilization and embryo transfer, which separate the act of intercourse from the processes of reproduction, are condemned for shattering the integrity of reproduction. "Fertilization achieved outside the bodies of the couple remains by this very fact deprived of the meanings and the values which are expressed in the language of the body and in the union of human persons.... In homologous IVF and ET, therefore, even if it is considered in the context of 'de facto' existing sexual relations, the generation of the human person is objectively deprived of its proper perfection: namely, that of being the result and fruit of a conjugal act in which the spouses can become 'cooperators with God for giving life to a new persons"' ([4], pp. 28, 30). This emphasis on not sundering the unitive and the procreative elements of reproduction has also become the basis for recent Roman Catholic condemnations of contraception, as with Pope Paul VI's 1968 encyclical letter, *Humanae Vitae*.[8]

The Congregation for the Doctrine of the Faith also criticizes the new reproductive technologies for objectifying the child to be. "He cannot be desired or conceived as the product of an intervention of medical or biological techniques; that would be equivalent to reducing him to an object of scientific technology" ([4], p. 28). This highlights a theme of criticism in *Donum Vitae*, which appears to spring not just from a spiritual understanding of human reproduction, but from an account of the normatively humane life.[9]

The document *Donum Vitae* in addition criticizes the new reproductive technologies for making the child tantamount to a product rather than accepting the child as a gift of God. "Every human being is always to be accepted as a gift and blessing of God" ([4], p. 23). Here, the Congregation for the Doctrine of the Faith expresses a concern about the ways in which children will be regarded, if reproduction becomes a triumph of persons over human nature. The Congregation disapproves of persons taking control over their reproduction, as, for example, persons have taken charge through medicine of other areas of human physiology. Finally, the document reiterates the well-known Roman Catholic criticism of abortion, focused here on the destruction or discarding of extra or seemingly defective preembryos.

The force of the Roman Catholic position in all of this is not to disregard the goods of this life, but to see them in terms of the next. From the spiritual perspective of a religion with transcendental commitments, everything looks different. As a result, the Roman Catholic position regarding the use of new reproductive technology cannot be assessed by focusing only on the secularly assessable benefits and harms associated with the human reproductive technologies. A believing Roman Catholic will bring to the circumstances at hand a quite different interpretative schema for calculating harms and benefits than will an atheist. Their schedules for the assessment of costs and benefits will be incommensurable because of the believers' introduction of transcendent considerations. There appears to be an insurmountable barrier to consensus formation about the moral significance of the new reproductive technologies.

III. SACRED SEX VERSUS YUPPIE SEX

For the religious person who does not live within a Roman Catholic religious perspective, the considerations advanced by the Congregation for the Doctrine of the Faith will in many areas appear wrongheaded. But for the person who lives outside of any religious tradition or traditional understanding of human life, the Congregation's position will appear bizarre and exotic, somewhat like the tales of mutilative religious rites of primitive peoples. Indeed, in an increasingly secular world, where major religious practices have become polite charades of a once-vivid religious past, the very idea that men and women in a highly technological society would array their lives around transcendent spiritual goals appears for many to be nearly incomprehensible. The once regnant Christianity of the West has become a collection of sects and cults.

It is not just that religion has been marginalized in Western societies. Equally significant is the loss of a normative understanding of being human and of human nature. Within the framework of modern science, human nature is the accidental outcome of the results of spontaneous mutation, past selective pressures, random catastrophic events, genetic drift, and other biological, chemical, and physical constraints and forces. There is nothing special about the way humans are constructed. The general anatomy and physiology of the human body is the best that selection has been able to provide, and this is often not good enough. There are many elements of human nature that are still poorly designed to meet the interests and goals of persons (e.g., osteoporosis consequent upon menopause). Moreover, selection insofar as it has been successful has for the most part adapted us for an environment in which we no longer live (e.g., we have not lived in contemporary urban environments long enough to become well adapted to this niche).

Within this secular perspective, one can with good warrant regard nature as dominating, curtailing, and circumscribing the wishes, goals, and projects of persons. However, through medicine and the biomedical sciences (including genetic engineering) persons can (and in the future ever more will be able to) alter, redesign, and refashion their human nature. For example, nature "naturally" aborts most, but not all, defective pre-embryos. It becomes quite "natural" for persons to augment the efficiency of nature, to realize the quite straightforward goal of having children with as few mental and physical handicaps as possible. Human reproductive capacities become one among the many human functions to be controlled and directed by human goals and projects. In particular, contraception, sterilization, prenatal diagnosis and abortion, as well as third-party technologically assisted forms of human reproduction, offer persons ways of escaping the domination and tyranny of human nature.[10] Again, because the particular character of human nature is merely the factual outcome of biological processes, the constraints human nature sets on the wishes, goals, and projects of persons appear as surd, irrational, and pointless (at least outside of a special interpretative perspective). After all, if there is no Designer, there is no Providence. Moreover, there is no redemptive transcendent significance of accepting pain, suffering, and the lot that nature gives. For the contemporary man and woman, there is no "Nature", but only "nature" as a collection of causal forces to be controlled and directed as one controls and directs wild streams with dams and levees.

A technologically mediated dialectic develops between persons and their human nature. Individuals as rational beings, as persons, are able to examine themselves critically and define their own goals and objectives. As if they were gods or goddesses, men and women as rational entities can envisage indefinite lifespans, free of illness, disability, and distress. They can objectify those elements of their human nature that limit lifespan and occasion illness, disability, and distress. They can then envisage how technologically to set those limitations aside. With technological success, the critical stance of the self-conscious person is augmented with confidence that the project of refashioning human nature is feasible.

This project of personalizing human nature (i.e., ensuring that human nature supports the goals and projects of persons) comes to encompass the project of giving birth to children innocent of illness, disability, and distress. It embraces both technologies to hinder reproduction, as well as technologies to enhance reproductive capacities, so that projects of having (or not having) progeny can be realized despite the hindrances of nature. The domination of nature by persons becomes the liberation of persons from nature. Rather than being defenselessly at the mercy of plagues, illness, over-population, and sterility, persons can envisage through medical technology vaccines against illnesses, treatments for diseases and disabilities, contraceptives to prevent over-population and unwanted children, as well as ways of restoring fertility.

This secular language of domination of human nature in order to liberate humans as persons contrasts dramatically with the attitudes expressed in *Donum Vitae*. The document rejects using genetic engineering to ensure that one has children with the health and capacities one values. "Certain attempts to influence chromosomal or genetic inheritance are not therapeutic but are aimed at producing human beings selected according to sex or other predetermined qualities. These manipulations are contrary to the personal dignity of the human being and his or her integrity and identity. Therefore in no way can they be justified on the grounds of possible beneficial consequences for future humanity" ([4], pp. 19–20). The document regards the attempt to domesticate nature in the case of human reproductive capacities so that it conforms to the wishes and projects of persons to have a child, to be an illicit domination of nature. "Homologous IVF and ET is brought about outside the bodies of the couple through actions of third parties whose competence and technical activity determine the success of the procedure. Such fertilization entrusts the life and identity of the embryo into the power of doctors and biologists and establishes the domination of technology over the origin and destiny of the human person. Such a relationship of domination is in itself contrary to the dignity and equality that must be common to parents and children" ([4], p. 30). One finds here a strong contrast between a view that accepts God's design in Providence and/or Nature in the conception of children and a view that turns to nature in order to make it serve the goals of persons, namely, to have healthy children of their own.[11]

Those who live outside of religious or other particular normative traditions have no reason to regard the reproductive, social and recreational elements of sexuality, save in mundane, secular terms. For them, sex is obviously for fun, relaxation, companionship, and occasionally for having children. Sex is good and enjoyable, but without transcendent meaning. In contrast, for the person who lives within a traditional religious community, everything has transcendent significance. Sex, celebration, suffering, and death, all are given a meaning that transcends the present and the particular individual. For the individual in a religious tradition, sex becomes a gift of God, an occasion for the blessing of the marriage bed, a mitzvah on the Sabbath, and an act to be avoided on particular days or times for ritual purposes. Sex, as all elements of life in a traditional community, is given a place and purpose, such that those from outside the community must ask about its significance and the circumstances under which it may take place.

In contrast, in the secular, international society of individuals who live outside of any robust religious or traditional interpretive framework, the world is without transcendent significance. There is no deep explanation to be given regarding the significance and place of sex, suffering, and death. They are as one finds them and as one experiences them. Within this secular society that spans from Brazilia to Montreal, from Toronto to Tokyo, from Berlin to Adelaide, from Buenos Aires to Paris, its individuals understand each other and share immanent concerns, interests, pleasures, conflicts, and life projects. They recognize each other as individuals pursuing the obvious goods of a world-wide, post-industrial, technological society. They are people of everywhere and nowhere, for whom technological interventions that enabled them better to control their reproductive capacities are obvious goods.

Alasdair MacIntyre has termed the general secular languages of this contemporary secular world the internationalized languages of modernity [18]. MacIntyre describes the individuals who speak these languages as rootless cosmopolitans. In this, he identifies what somewhat tendentiously might be called the international yuppie culture, marked by a striving after the good things of this life, unhindered by special transcendent constraints or considerations.[12] "... [T]he social and cultural condition of those who speak that kind of language, a certain type of rootless cosmopolitanism, the condition of those who aspiring to be at home anywhere – except that is, of course, in what they regard as the backward, outmoded, undeveloped cultures of traditions – are therefore in an important way citizens of nowhere is also ideal-typical. It is the fate toward which modernity moves precisely insofar as it successfully modernizes itself and others by emancipating itself from social, cultural, and linguistic particularity and so from tradition" ([18], p. 388).

In this secular language, sex, as all else, has an instrumental value. Reproductive sex, in particular, serves the goal of producing a healthy, happy child. But, the having of children is itself bereft of transcendent significance. For the cosmopolitan, there is no religious or traditional significance in having a child. One may have a child in order to experience child-bearing and child-rearing. One may have a child because one is looking forward to companionship in old age. But there are no transcendent reasons to have children, as exist within traditional religious communities where because of Divine requirement sex may be enjoyed only within marriage, where marriage is valid only when there is an intention to produce children, and where the production of children fulfills a Divine purpose ([19], p. 14). Even non-religious traditional societies give an account of childbearing that transcends the immediate individuals involved by underscoring the life of the clan or of the cultural or ethnic group. In the secular context, should one want a child and technological assistance is necessary, the only considerations are the feasibility of the procedures, the cost of the procedures, and the consent of those involved and whether having a child fits within one's life plans. The significant decrease in reproductive rates among groups that embrace the mores of cosmopolitans may indicate how often the project of childbearing recommends itself in purely secular terms.[13]

IV. WHY CONSENSUS IS HARD TO ACHIEVE

When it comes to the moral significance of third-party technologically assisted reproduction, cosmopolitans do not share a consensus with those who live their lives within the embrace of the traditional beliefs of the Catholic Church (or most other religious groups). The cosmopolitan will be interested in safety, efficacy, and costs.[14] The cosmopolitan will be concerned that the side effects of the treatments are well known and well disclosed, but will have no reason for thinking that human reproductive technologies are wrong in themselves.[15] The debate, the controversy, the dispute between the religious believer and the secular cosmopolitan does not appear amenable to resolution.

The irresolvability of this controversy is a part of a more fundamental problem confronting contemporary moral reflection and public policy. The history of the West can somewhat procrusteanly be seen as a passage from its ancient period when, under a polytheistic metaphor, numerous visions of the good life were entertained against the background of substantial philosophical skepticism. This period was succeeded by over a thousand years of effective Christian hegemony in which a monotheistic metaphor for moral reflection and political theory was imposed on the West. Not only was there a robust faith in a single God, but a robust faith in reason's capacity to discover a univocal moral account.[16] The modern age ushered in by the Renaissance attempted to secure a canonical philosophical account of morality and political theory without reliance on the traditional religion of the West. Modernity in this sense has been the attempt to have the morality and politics of Christian monotheism without belief in God.

The difficulty is that this project is essentially flawed. Given general agreement about certain epistemic values and about ceteris paribus conditions, one can fashion a common account of reality. The modern project of science and technology has triumphed. Moreover, scientific dissenters have generally been at liberty to go their own ways with alternative sciences (e.g., astrology still flourishes). However, the modern projects of providing a rational foundation for a content-full morality and public policy seem unfeasible. In Section I, the example was given of the difficulty of providing a morally canonical ranking of equality, liberty, security, and prosperity. Depending on the rank one gives to these social desiderata, it was noted, one will live in a quite different moral world. But unlike science, where dissenters are rarely coerced into abandoning their knowledge claims but are rather usually only banished from public funding, governments even in democratic, secular pluralist societies attempt to enforce a particular morality.

The project of modernity has included the attempt to secure a justification for governmental coercion by deriving a morality from reason itself, from reflection on a calculus of pains and pleasures or from some other device meant to show that rational individuals should endorse a particular morality, along with its political implications. A government created in accord with this morality should then have the warrant of rationality to use force. Anyone who rejected the morality would, ipso facto, be acting irrationally. Anyone who understood morality as a rational endeavor would realize that the coercing government was acting rationally, that is, morally. Moreover, the coerced persons could not object on moral grounds. In addition, coerced persons would not be alienated in the process of coercion but receive the imposition of their true rational self, of what they ought to affirm.

The difficulty is in discovering in reason or in nature a canonical vision or ranking or portrayal of human goods that can contentfully guide public policy with moral authority. In order to know how to rank social desiderata such as equality, liberty, prosperity, and security, one must already possess the correct moral sense, correct thin theory of the good, or correct moral vision. In order to decide which account has the better consequences, one will need first to know how one ought to compare and weight equality, liberty, prosperity, and security consequences. Appeals to preferences will not be definitive either, unless one knows how to rank impulsive preferences with well considered preferences, present preferences with future preferences. One must be able to decide when a state intervention to solve a coordination problem will cost more in liberty values than it gains from coordination in respect to other values. In order to select the correct moral vision, account of consequences, account of preferences, one must already know the answer to the question, which is to say that the question is unanswerable.

What has been said about ranking equality, liberty, prosperity, and security can be said as well about different views of reproductive integrity or relationships to human nature. Moral consensus with regard to the morality of human reproductive technologies is unattainable in general secular rational terms because there is no uncontroversial account of moral reasoning or of practical reasoning to which one can appeal for direction.

V. THE POST-MODERN PREDICAMENT

If one cannot through reason definitively establish a content-full moral account, one cannot appeal to reason's authority to endorse a particular moral vision. The hope of the modern age to be able on the basis of reason to establish a canonical and content-full moral account that can authorize a content-full public policy is vain. Even without introducing transcendent considerations, individuals with different rankings of more mundane concerns can have irresolveable moral disputes. However, when one adds transcendent religious moral beliefs to the divergence of possible non-religious moral viewpoints, the moral world fragments into even more profound differences. One finds the contemporary post-modern world in which the polytheistic metaphor has been restored and in which numerous moral views contend.

The seemingly endless moral controversies regarding content-full moral issues such as the proper use of human reproductive technologies do not lead to the total collapse of secular ethics. Even if God is silent, and even if one cannot by appeal to reason alone ground an authoritative moral viewpoint, still, insofar as one is interested in resolving moral issues with common authority, one can establish a moral framework justified in terms of mutual consent. Within such a framework, no one is a moral authority in the sense of being able to show in general secular terms which content-full moral vision ought to be endorsed. However, individuals can be in moral authority in the sense of being chosen by some to act on their behalf [10]. Secular moral authority takes on the very straightforward, non-metaphysical sense of authorization by actual individuals to do things for them. A moral framework is thus not discovered by reason, but created by common consent.

The difficulty is that such consent must be unanimous with regard to collaboration in any project. However, this difficulty is not as insurmountable as one might at first envisage. To begin with, one must remember that the point of departure in moral reflection is the interest in finding a basis for showing why individuals ought or ought not to act in particular ways, and for showing why, under particular circumstances, they ought to recognize the moral authority of coercive political power. Traditionally, moral authority has been derived from some Divine appeal within a particular moral tradition. The West attempted to supplant appeals to the Divine by an appeal to reason. If one is in a secular pluralist context, an appeal to Divine authority will not ground moral claims that individuals can commonly endorse. It is not simply that such an appeal will not motivate moral agents to comply, it is that without the special premises derived from a religious perspective or from a particular moral tradition there will not be grounds for justifying the appeal as rationally conclusive.

The appeal to reason was attractive in that it seemed to offer a way of stepping outside of all traditions and still coming to know concretely what one ought to do. The goal was to derive moral authority from reason. As was noted above, had this appeal been successful, one could have resolved rational questions about morality for rational agents. If this is not feasible, one stands on the brink of nihilism unless one either converts to a particular moral or religious tradition and/or finds a different strategy for securing general secular moral discourse. Such a strategy is available through recognizing what it is to resolve moral controversies without recourse primarily to force. If one is interested in resolving moral controversies without recourse to what will in general secular terms appear to be an appeal to force, and if God is silent with respect to the dispute, and if reason cannot discover a concrete morality, then there is still the possibility, on the basis of mutual respect and mutual agreement, to maintain a general secular moral language and to act with general secular moral authority.

Moral authority in this circumstance is derived from the general practice of resolving issues by mutual consent or from particular agreements. In the first case, one is authorized to do all that is necessary to maintain the practice within the confines of mutual respect and one is forbidden to do that which is in contradiction with the practice. Thus, one is forbidden to engage in robbery, rape, and murder, and allowed to employ punitive and defensive force against those who do. Moreover, those who wish to use unconsented-to force on the innocent have no general secularly justifiable basis to protest, for the only general secular moral language left is that of mutual respect. Since the practice of secularly resolving issues peaceably does not claim a religious or transcendent justification, one will not be able to say, in general secular terms and *sub specie aeternitatis*, that it is good to resolve issues peaceably. The practice of secular ethics is simply a practice within which one may enter for many reasons, none of which is amenable to a general, secular, philosophical justification.[17]

If one lives in a pluralist context, and if God is silent, and if reason cannot discover a content-full morality, one can still engage in a wide range of communal activities with general secular moral approbation. One can protect individuals against being used without consent. One can enforce all recorded contracts. Moreover, one can establish agreements with regard to the use of communal resources. However, without explicit consent it will be impossible to justify enforcing a particular content-full morality. In short, one finds a general secular justification for many contemporary ethical and political practices. One sees why the practice of free and informed consent is so salient in health care: if one cannot discover what patients ought to do, patients and physicians must agree together what they will do jointly. One sees as well why limited democracies are so salient, not simply because of a value assigned to liberty, but because they afford the only plausible source of political authority, when appeals to God or content-full moral reason are impossible. It becomes as implausible that majorities should rule by Divine right or right of reason, as it is that enlightened despots should rule by Divine right or right of reason. But if there are limits on the authority of majorities to control the beliefs and actions of the members of a society, rights to privacy become salient as well.

VI. IF YOU CAN LIVE WITH RIGHTS TO PRIVACY, YOU CAN LIVE WITHOUT A MORAL CONSENSUS REGARDING THE USE OF HUMAN REPRODUCTIVE TECHNOLOGIES

Rights to privacy, a concept from American law, provide the key to the problem of developing a moral consensus with regard to the use of human reproductive technology.[18] As has already become clear, it will not be possible to discover a normative basis for a concrete moral consensus in a secular pluralist society, and it is very unlikely that such a consensus will develop with regard to the use of the new human reproductive technologies. Moreover, if one were to believe that such a consensus could with authority be fashioned, it would lead, at least from a general secular point of view, to an unauthorized imposition of a particular moral view on unconsenting minorities. However, if one abandons the hope of a general conversion to a particular religious view, or to the rational discovery of a particular concrete moral perspective, and if one recognizes the limitations of moral reason, one can altogether abandon the pursuit of a concrete consensus with respect to the use of the human reproductive technologies.

This is not to say that one should in all areas lose interest in consensus. After all, one will have moral grounds for endorsing a consensus (with agreement approaching unanimity as far as possible) regarding the moral obligation not to use others without their consent. Moreover, insofar as one establishes by various contracts agreements to distribute common resources in majoritarian fashions, it will be useful to have as much common agreement as is possible in order to minimize social discord.

In addition, there will be many areas in which the use of commissions and fora for public discussion will allow a significant majority to be established, so that particular projects can be more easily undertaken. However, in these last cases one must distinguish a significant majority that allows for the realization of a project from a majority treated as an morally authoritative consensus that would put minority views in danger of being suppressed by coercive social force.[19]

If there is a consensus to be pursued, it should be with regard to the post-modern condition: there are a plurality of concrete moral perspectives and traditions, and secular reason is without content; moreover, it is possible to resolve issues peaceably. In light of this consensus, one can then attempt to delineate the possibility for communal moral authority with respect to human reproductive technologies. There will be a significant scope of issues with respect to fraud, failure to make proper disclosures, etc., to which societies should turn to protect individuals, whatever their moral interests or concerns with reproduction might be. However, individuals should be free to use or not use these technologies, as they wish, as long as those who disagree are not constrained to collaborate with them. Finally, those who disagree should be at liberty peaceably to announce the damnation, particular and general, of all who use these technologies. In the absence of the possibility of a concrete consensus with regard to the moral significance of the human reproductive technologies, there should be freedom to go to hell as one wants, and to damn those who appear headed in that direction.

Center for Ethics, Medicine and Public Issues
Baylor College of Medicine
Houston, Texas, USA

NOTES

[1] There is no reason to believe that social harmony and concord will be valued in themselves by all individuals or valued without exception. There are many who value hating their neighbors and who yearn for an opportunity for belligerence and strife. There is no question that some amount of consensus will be valued by all who are not loners in their endeavors of strife and hostility. For example, if one wishes to make a practice of pillaging the neighboring village, one may value pillaging efficiently and therefore seek some amount of consensus with one's co-pillagers. See, for example, an account of the life of the Yanomamö [3]. The point is that the vision of the good life as one of general social harmony and concord is not one that will necessarily claim

the affirmation of all rational beings, unless one can show that rationality commits one to endorsing universal social harmony and concord. Assigning a high value to social harmony and/or consensus would seem to involve a choice of one among many possible visions of the good life.

[2] That consensus can be useful is not a controversial point. Nor does it directly help us come to terms with framing policy with respect to reproductive technology. After all, those who disagree on important moral grounds will hold that they have good reasons for not being part of a general consensus.

[3] One might think of the praise heaped on the Pax Romana by the Christian poet Aurelius Clemens Prudentius (A.D. 348–c.410), who extolled Rome as having secured the material conditions for the spread of Christianity. "Rome without peace finds no favour with Thee; and it is the supremacy of Rome, keeping down disorders here or there by the awe of her sovereignty, that secures the peace, so that Thou hast pleasure in it" ([25], p. 57).

[4] The development of the concept of a *ius gentium*, in contrast with a *ius civile*, enabled the Romans to deal better with diverse populations encompassed within the Empire. Gaius, in his *Institutes*, for example, speaks of the common law of mankind, the "ius gentium, quasi quo iure omnes gentes" (*Institutes of Gaius* I.1). The *ius gentium* came to be distinguished from the *ius naturale* or the *lex naturae*, the latter concept being imported into Roman reflection by such thinkers as Cicero, who at times uses the phrase *lex mundi*. The *ius gentium* came also to be understood as a generalization from the legal systems of other nations, as well as an expression of the existence of the *ius naturale*. "Quod vero naturalis ratio inter omnes homines constituit, id apud omnes populos peraeque custoditur vocaturque jus gentium, quasi quo jure omnes gentes utuntur." *Institutes of Justinian*, Lib. I, Tit. II.1. See, also [5]; [27].

[5] "Consensus fidelium est certum Traditionis et fidei Ecclesiae criterium" ([28], § 1139). "Sententia communis" was taken to be a synonym for "consensus fidelium". For a matter to be identified as a part of the authority of tradition, it must be present certainly (certus) and clearly (clarus) and involve a matter of faith and morals (res fidei et morum).

[6] St. Thomas Aquinas, for example, argued that, all else being equal, masturbation should be considered a greater evil than adultery, because masturbation violates the very laws of God in nature. Masturbation directly injures God while adultery directly injures one's neighbor (Summa Theologica, II–II, 153–154). Generally, this argument against masturbation does not exist in the Orthodox Catholic Church. Masturbation is considered a sin, but carries only the penance of 40 days excommunication ([20], p. 936). In contrast, fornication carries an excommunication of two years ([20], p. 939).

[7] The Greek Orthodox Church does not appear to condemn the traditional fertility workup or artificial insemination from the husband ([6], p. 16). In contrast, the Ukrainian Autocephalous Orthodox Church in Texas supports a forty days excommunication for the traditional fertility work-up for males (requiring masturbation). Here the imposition of penance appears primarily to serve the goal of emphasizing that one should mourn that one has departed from an ideal conjugal relation in order to have a child. The ideal would be (1) to understand that "the gift of sexuality reflected in all of nature is from our Great God Who hath revealed Himself unto us. As with all capacities in our human nature, the capacity for sexuality can be mundane or consecrated" ([19], p. 11), and (2) to accept the lot God has given one, that is, to be fertile or infertile ([19], p. 13). All of this is elaborated in terms of a view of marriage as blessed by God, a point

made somewhat strongly in the old Apostolic Canon LI. "If any Bishop, or Presbyter, or Deacon, or anyone at all on the sacerdotal list, abstains from marriage, or meat, or wine, not as a matter of mortification, but out of an abhorrence thereof, forgetting that all things are exceedingly good, and that God made man male and female, and blasphemously misrepresenting God's work of creation, either let him mend his ways or let him be deposed from office and expelled from the Church. Let a layman be treated similarly" ([20], p. 91). The theme in this reflection on AIH is that certain activities aimed at assisting reproduction substantially distract from a life consecrated to God.

[8] It is interesting to note that Orthodox Catholicism does not forbid the use of contraception. "When as a temporary measure to delay childbearing or when couples have had the full number of children which is desired based upon health and economic factors, fertilization prevention may be practiced as a concession to the weakness of the flesh but should be mourned as conceding to human frailty" ([19], p. 14). Notable here is the absence of Scholastic, Thomistic-Aristotelian arguments regarding contraception. Instead, one finds a spiritual understanding of the marriage union and an interest in underscoring transcendent concerns.

[9] I have in mind the normative concept of living well, which the Romans described as living *humaniter*. The pagan Roman understanding of the ideally human, that which is *humanus*, appears recast here in Christian terms, as a result of the long passage of the *studia humanitatis* through the *studia divinitatis*. See, e.g., [15].

[10] Human nature in general secular terms is not a normative concept. It identifies the particular anatomical, physiological, genetic, and psychological characteristics that mark members of the species homo sapiens. Medicine in contrast employs various normative concepts of human nature. But none of these can be established as canonical on the basis of general secular considerations. See [9].

[11] I do not mean to deny the existence of a general suspicion on the part of many regarding the use of technology and regarding the domination and exploitation of nature. Such critiques of contemporary technology are rarely so radical as to suggest that humans should go back to being hunter-gatherers, innocent of the benefits of vaccines and effective contraception.

[12] I use the term yuppie here in a broad sense to include individuals of all ages who aspire after the general secular goods of an international urban civilization. By noting that the paradigmatic yuppie is unconstrained by transcendent considerations, I do not mean to suggest that my stereotypical yuppie is untouched by moral considerations. It is only that the moral considerations will be predominantly, if not exclusively, instrumental or utilitarian in character. *The Yuppie Handbook*, for example, gives no guidance with regard to attending religious services, save in order humorously to indicate the non-traditional character of yuppie religion. "Yuppies have their own form of organized religion. They invariably worship at the altar of self-improvement" ([24], p. 69). Moreover, the chapter on "The Yuppie Wedding" indicates only in passing that one should "meet with chaplain" (p. 93). In this I take *The Yuppie Handbook*, though it is offered as a satire, to provide a good sociological account of the relative importance of transcendent concerns in the lives of yuppies. In short, I use the term 'yuppie' as a somewhat tendentious way of presenting the moral significance of Alasdair MacIntyre's cosmopolitan. MacIntyre himself does not use the term "cosmopolitan", but rather speaks of those who live a "rootless cosmopolitanism". It must also be stressed that my characterization of yuppies is not meant to speak to the numerous young upwardly

mobile professionals who are devout members of particular religions or who live well and successfully within the embrace of a content-full tradition. Only some of the young upwardly mobile professionals are yuppies as I advance the term.

[13] One should note that traditional groups, such as Orthodox Jews and Mormons, continue to have high reproductive rates, even though the members of these groups function very well within highly technological societies.

[14] There are a number of groups with strong moral concerns regarding the use of human reproductive technologies who are neither a part of religious communities nor a part of traditional moral communities. I have in mind here individuals who regard the human reproductive technologies as morally suspect out of sympathy for an atechnological relationship to nature in general, or because they hold that this technology tends to subjugate or dominate women. What is said in the body of the text with regard to religious and traditional moral groups applies with respect to these groups as well, for they, too, bring to reality special assumptions and moral premises.

[15] There will not be general secular grounds for rejecting reproductive technologies, which involve the destruction of embryos, in that there will be no general secular grounds to show that embryos have an important, intrinsic moral status. See [9].

[16] Much of the success of modern atheism has been laid at the feet of Roman Catholicism's attempt to justify its faith in terms of reason. See [2]. Roman Catholicism has the peculiar distinction of having declared as a matter of faith that one can on the basis of reason alone prove the existence of God. "If anyone shall have said that it is not possible to know certainly the one and true God who is our Lord and Creator by the light of natural human reason through those things that have been made, may he be anathema." Constitutio dogmatica de fide catholica, Canones, II. De revelatione, 1, from the Fourth Session of the Vatican Council, 24 April 1870 (my translation).

[17] What I have provided is tantamount to a transcendental justification for a contentless secular procedural morality. Like most transcendental arguments, this involves laying out the grammar or conditions for a human practice that is so central to the life of persons that it is nearly unavoidable, that it is relatively a priori. However, as a transcendental, not a metaphysical, argument, I do not attempt to show why one must enter the practice or how the practice is embedded in the conditions of reality as such.

One may enter this practice for numerous reasons, including religious reasons. One might be of the view that the general secular government should not be in the business of coercing individuals into morally correct behavior, if the behavior does not involve unconsenting participants, or into converting to the proper concrete moral perspective. See, for instance, the CXIXth canon of the Council of Carthage (A.D. 418/419), which was later affirmed by the sixth and seventh Ecumenical Councils: "There has been given a law whereby each and every person may by free choice undertake the exercise of Christianhood" ([20], p. 673).

[18] The history of the American concept of rights to privacy is a complex one. On the one hand, the phrase has a substantial history in American tort law, in which Samuel Warren and Louis Brandeis played a large role [29]. For this essay, a second contribution of Brandeis, namely, to American constitutional law, is more important. In the holdings of the American Supreme Court in the 20th century, opinions have been written indicating that individuals have transferred some but not all authority over themselves to their government. The areas where transfer of authority has not, or only incompletely, taken place, remain as areas of free personal choice, areas where citizens

maintain rights to privacy. As Brandeis, while a Supreme Court Justice, advanced the notion, "The makers of our Constitution undertook to secure conditions favorable to the pursuit of happiness. They recognized the significance of man's spiritual nature, of his feelings and of his intellect. They knew that only a part of the pain, pleasure, and satisfactions of life are to be found in material things. They sought to protect Americans in their beliefs, their thoughts, their emotions and their sensations. They conferred, as against the Government, the right to be let alone – the most comprehensive of rights and the right most valued by civilized men" ([22], 478). This notion of a right to privacy reflects an old Germanic view that governments are created by the consent of individuals and that governments have only limited authority over their citizens ([16], pp. 24–5). This American constitutional view has been developed recently with regard to issues of contraception and abortion ([13]; [7]; [26]). See, also, [21]; [23]; [1]. The constitutional standing of rights to privacy in American constitutional law are currently a matter of significant controversy in the United States. The foundational moral issues remain unchanged.

[19] This is not the place to develop a general account of political theory as it must be understood in a secular pluralist context within which appeals to God or to concrete moral reason fail. However, see [9].

BIBLIOGRAPHY

1. Barnett, R.E. (ed.): 1989, *The Rights Retained by the People*, George Mason University Press, Fairfax, Virginia.
2. Buckley, M.J.: 1987, *At the Origins of Modern Atheism*, Yale University Press, New Haven.
3. Chagnon, N.A.: 1977, *Yanomamö: The Fierce People*, 2nd ed., Rinehart and Winston, New York.
4. Congregation for the Doctrine of the Faith: 1987, *Instruction on Respect for Human Life in its Origin and on the Dignity of Procreation <Donum Vitae>*, Vatican City.
5. de Zulueta, F. (ed.): 1975, *The Institutes of Gaius,*, Clarendon Press, Oxford.
6. Department of Church and Society, Greek Orthodox Archdiocese of North and South America: (undated), *Statements on Moral and Social Concerns*, New York.
7. Eisenstadt v. Baird, 405 U.S. 438, 92 S.Ct. 1029, 31 L.Ed.2d 349 (1972).
8. Eliot, T.S.: 1958, *The Complete Poems and Plays*, Harcourt, Brace, New York.
9. Engelhardt, H.T., Jr.: 1986, *The Foundations of Bioethics*, Oxford University Press, New York.
10. Flathman, R.E.: 1982, 'Power, Authority, and Rights in the Practice of Medicine', in G.J. Agich (ed.), *Responsibility in Health Care*, D.Reidel, Dordrecht.
11. Gauthier, D.: 1986, *Morals by Agreement*, Clarendon Press, Oxford.
12. Gauthier, D.: 1987, 'Taming Leviathan', *Philosophy & Public Affairs* 15 (Summer), 280–298.
13. Griswold v. Connecticut, 381 U.S. 479, 85 S.Ct. 1678, 14 L.Ed.2d 510 (1965).
14. Hampton, J.: 1986, *Hobbes and the Social Contract Tradition*, Cambridge University Press, Cambridge.

15. Jaeger, W.: 1943, *Humanism and Theology*, Marquette University, Milwaukee.
16. Lea, H.C.: 1973, *Torture*, University of Pennsylvania Press, Philadelphia.
17. Lewis, C.T. and Short, C. (eds.): 1980, *A Latin Dictionary*, Clarendon Press, Oxford.
18. MacIntyre, A.: 1988, *Whose Justice? Which Rationality?*, University of Notre Dame Press, Notre Dame, Indiana.
19. Makarios, Bishop Vladika: 1989, 'Human Sexuality in the 1980's', *Orthodox Outreach* 12 (October-December), 11–15.
20. Nicodemus and Agapius: 1983, *The Rudder of the Orthodox Catholic Church*, Orthodox Christian Educational Society, Chicago.
21. O'Brien, D.M.: 1979, *Privacy, Law, and Public Policy*, Praeger, New York.
22. Olmstead v. United States, 277 U.S. 438 (1928).
23. Patterson, B.B.: 1955, *The Forgotten Ninth Amendment*, Bobbs-Merrill, Indianapolis.
24. Piesman, M. and Hartley, M.: 1984, *The Yuppie Handbook*, Pocket Books, New York.
25. Prudentius: 1953, 'A Reply to the Address of Symmachus (Contra Orationem Symmachi)', in *Prudentius*, vol. II, trans. H.J. Thomson, Harvard University Press, Cambridge, Mass.
26. Roe v. Wade, 410 U.S. 113, 93 S.Ct. 705, 35 L.Ed.2d 147 (1973).
27. Sandars, T.C. (ed.): 1970, *The Institutes of Justinian*, Greenwood Press, Westport, Conn.
28. Tanquerey, Ad.: 1937, *Synopsis Theologiae Dogmaticae Fundamentalis*, Society of St. John the Evangelist, Paris.
29. Warren, S. and Brandeis, L.: 1890, 'The Right to Privacy', *Harvard Law Review* 4, 193–220.

THE CONCEPT OF MORAL CONSENSUS

*Philosophical Reflections**

A moral philosopher examining social reality finds himself confronted with a remarkable fact: while the occurence of consensus and agreement regarding central moral questions seems to be decreasing in society, *the concepts* of consensus and agreement are becoming increasingly attractive for ethical theory. In Anglo-Saxon moral philosophy the key role played by the concepts of consensus and agreement is chiefly attributed to *contractualism*. The renaissance of the contractualist approach can be traced to John Rawls, according to whom the basic principles of justice achieve moral viability due to the fact that they are established consensually in the original position and under certain conditions by the persons concerned. Following Rawls' thesis "that the argument for the principles of justice should proceed from some consensus" ([14], p. 581), other authors have expanded upon the foundational function of consensus, beyond the principles of justice. According to T.M. Scanlon, for example, the validity of each moral principle must be attributed to a "hypothetical agreement" ([15], p. 44) which is entered into voluntarily by free and rational persons. The concept of consensus also plays a prominent role within German philosophy, most notably in the *Diskursethik*, developed by Karl-Otto Apel and Jürgen Habermas. This approach differs from contractualism through the two phases of its foundation program. The first phase of this program aims at the transcendental-pragmatic foundation of a criterion for moral rightness: According to the *Diskursethik* a norm only has a claim to validity if all those potentially affected by that norm reach (or would reach) agreement as participants in a practical discourse that the norm is valid ([9], p. 76). On the basis of this principle concrete moral discourses are then required in order to discuss and consensually confirm (or not) the validity of individual material norms.

This role played by "consensus" as a basic ethical concept is new and

Kurt Bayertz (ed.), The Concept of Moral Consensus, 41–57.

validity of individual material norms.

This role played by "consensus" as a basic ethical concept is new and requires interpretation. The considerations which follow aim to provide a sketch of such an interpretation. First of all, I will investigate the philosophical and historical roots of the ethical concept of consensus, and will try to identify the philosophical problem to which it is offered as the solution. The second section will deal with the consequences for the concept of morality arising from the new central importance of "consensus". Thirdly and finally, I will attempt to define the implications which the concept of consensus has for practical ethics and its limitations as a solution for the moral problems surrounding human reproductive technologies.

I. FROM POLITICAL TO MORAL PHILOSOPHY

(1) The idea of consensus has a long tradition in ethical literature. As part of his discussion on the connection between human nature and pleasure Aristotle refers to consensus as evidence for the correctness of statements: a conviction shared by all human beings is correct and there is nothing more convincing to counter this argument (*Ethica Nicomachea,* X 2, 1173a, [1]). In a similar way consensus is regarded as a criterion for truth by other ancient world authors (e.g. Cicero *De Divinatione*, I, 1; I, 11; I, 84; [4], pp. 223, 235, 317). Yet it is not until the political philosophy of the 17th century that the concept acquires a key philosophical function. Political thinking in the modern age is confronted with a new problem, both historically and philosophically: how is it possible to justify the existence of the state and its power over the people when the latter are free and independent individuals? For the political philosophy of the ancient world (and the Middle Ages) the matter of justification of the state had not arisen in this form or this acutely. For Aristotle, for example, man is a natural "political being", that is, destined for a life within social structures and state institutions; the state is primary, and not the individuals (*Politica,* I 2, 1253a, [2]). The Modern Age, in contrast, sees man *as an individual*, that is, as a naturally free and unsocial being, only prepared to come together with others in larger social groups when placed under external pressure. Hobbes specifically emphasizes that "agreement" between social insects differs entirely from agreement between socialised human beings: "...the agreement of

these creatures is natural; that of men, is by covenant only, which is artificial..." ([10], p. 157). For Hobbes or Locke the state no longer precedes the individuals; on the contrary, individuals precede the state: the latter is an "artificial" outcome of the unification of free individuals.

It is obvious that this turnaround led to a drastic increase in the need for legitimation for the state. Society and state can no longer be presupposed as natural entities nor taken for granted as simple facts. A much greater burden of explanation and justification is incumbent upon political philosophy as far as the state is concerned. Against this background it becomes clear upon what the key role played by the concept of assent in political philosophy is based. The power of the state over individuals is justified by the idea of an – implicit or explicit – assent on the part of the individuals to this power.

> Men being, as has been said, by Nature, all free, equal and independent, no one can be put out of this Estate, and subjected to the Political Power of another, without his own *Consent*. The only way whereby any one devests himself of his Natural Liberty, and *puts on the bonds of Civil Society* is by agreeing with other Men to joyn and unite into a Community, for their comfortable, safe, and peaceable living one amongst another,... ([13], pp. 348f.).

This consideration that individual assent is the necessary condition for a political institution to be legitimate, and that therefore no legitimate state may exist without general consensus, becomes a basic consideration in modern political philosophy. From its origins in 17th century England it pervades classical German philosophy ([11], p. 205; [7], p. 174) through to contemporary theories of democracy ([6]; [3]). It should be emphasised here that in this case the concept of consensus acquires a much broader significance than in the ancient world reference to *consensus omnium*. For Aristotle or Cicero consensus functions as *indication* of the truth of a statement; a conviction is valid *if* it meets with general consent. In modern age political philosophy consensus is *the reason* underlying the legitimacy of the state; the state is legitimate *because* individuals give it their assent.

(2) This interpretation has obviously been adopted by the new theories of moral consensus. They confer from the field of politics to the field of morality the idea of a *constitutive* role of consensus and assent regarding the question of legitimacy. As political philosophers ask how "the bonds of Civil Society" can be legitimate when human beings are "free, equal and independent", so moral philosophers try to answer the question how the intersubjective obligatory nature of moral norms can be justified

when individuals are morally *autonomous*.

This moral autonomy is maintained in two directions. On the one hand, the concept of autonomy is directed against restrictions which could affect "from outside" the self-determination of the individual: through divine moral legislation, through the metaphysical roots of the principles of morality or through natural determinants of human decisions and actions. The theories of consensus deny the idea of a normative authority to which man has no access. Tracing all moral validity back to a free agreement between individuals makes man the creator of morality. Thus the moral theory of consensus assumes an understanding of moral truth, in the sense of a correspondence with natural facts or metaphysical laws. It is without doubt that this is one of the strengths of the theories of consensus. By making morality accessible to man they comply with the essential antimetaphysical feature of modern thinking. At the same time they express a basic characteristic of the moral situation of our time: the inevitability of "constructive" access to morality. Traditional morality is unable to answer (at least sufficiently) many of the questions arising with the advent of new technological possibilities in medicine. One of the most telling examples of this is the problem of technical intervention into human reproduction. We are forced to "construe" or "generate" a reproductive morality which is suitable for the technical options available today. The idea of consensus provides us with a criterion for the legitimacy of such "construed" norms: a moral norm to which everybody has agreed may legitimately oblige everybody.

There is room here merely to touch upon the fact that the ethical theories of consensus are thus a continuation of a central idea of classical political contractualism, and at the same time a radicalization in a way which would have been unimaginable for Hobbes or Locke. Hobbes' philosophy is particularly "constructivist", in the sense that it emphasizes constructive human action and views civil society and the state as a product of such an action: whereas animal societies are "natural", the state is "artificial". We have already seen that this view of political contractualism not only differs from ancient social philosophy, but also from contemporary efforts to give the state a superhuman basis. A fundamental tendency within modern thinking is that of turning as many naturally existing facts into the achievement of autonomous human beings. In the field of politics this program has been followed by classical contractualism very strictly. Yet neither Hobbes nor Locke would have been able to accept the idea of morality as "artificial", i.e., the result of human construction. With regard to morality, both are strict believers in the *Natural Law*. They believe the foundations of morality to be "natural" in the same way as the ties which bind in animal societies. In his *Essay Concerning Human Understanding* Locke does speak of a "secret and tacit consent" underlying views on virtue and vice within a particular society:

> Thus the measure of what is everywhere called and esteemed virtue and vice is this approbation or dislike, praise or blame, which, by a secret and tacit consent, establishes itself in the several societies, tribes, and clubs of men in the world: whereby several actions come to find credit or disgrace amongst them, according to the judgment, maxims, or fashions of that place ([12], p. 477).

This consent is not, however, arbitrary. True, there is a certain amount of leeway governing the approbation and dislike of moral rules, determined by the particular circumstances of place and time, so that "passes for vice in one country which is counted a virtue, or at least not vice, in another" ([12], p. 477). Yet at the same time this leeway is severely limited by Divine Law or the Law of Nature.

> And though perhaps, by the different temper, education, fashion, maxims, or interest of different sorts of men, it fell out, that what was thought praiseworthy in one place, escaped not censure in another; and so in different societies, virtues and vices were changed: yet, as to the main, they for the most part kept the same everywhere. For, since nothing can be more natural than to encourage with esteem and reputation that wherein every one finds his advantage, and to blame and discountenance the contrary; it is no wonder that esteem and discredit, virtue and vice, should, in a great measure, everywhere correspond with the unchangeable rule of right and wrong, which the law of God hath established ... whereby, even in the corruption of manners, the true boundaries of the law of nature, which ought to be the rule of virtue and vice, were pretty well preferred. ([12], pp. 478f.)

Thus, approbation is not *constitutive* for the validity of a moral rule; it is simply the means of carrying it through. To put it another way: approbation or dislike refer not to the moral, but merely the social validity of a rule. Modern theories of consensus, on the other hand, attribute not only the social but especially the *moral* validity of norms to approbation and consensus. That the theories of consensus have now extended the reach of man even to morality and interpret it as being an "artificial" social institution created by human beings in the way that Hobbes and Locke saw civil society and the state, may be interpreted as being as much an expression of continuity as of the epoch-making difference between the 17th and 20th centuries.

On the other hand the recent theories of moral consensus enforce "inwardly" the concept of autonomy. They seek to protect the self-determination of the individual against all threats which originate in society: in the violence inflicted by other persons or in the compulsion which can arise from existing structures of power. In the same way that political contractualism is aimed at illegitimizing the autocratic principle of power, the ethical theories of consensus are aimed against heteronomous morality. Human beings should themselves be able to determine the moral side of their lives, just as they determine their political fate in democratic societies. The idea of moral autonomy parallels that of political autonomy; and the idea of moral consensus corresponds with that of political consensus. In both cases consensus appears as the necessary medium in the imparting of the self-legislation of many individuals to a uniform will. Thus the moral theories of consensus may be viewed as a transferral of the principle of democracy to the field of ethics.

II. THE GOOD AND THE JUST

(3) This transferral comes up against a serious problem, however. The structural characteristic of modern democracy is not the agreement of all citizens but – on the contrary – the institutionalization of dissent, conflict, and controversy in the form of opposition, of freedom of the press, and of opinion. Modern democracy demands not consensus regarding political programs and goals, but acceptance of an institutional framework which on the one hand guarantees peaceful competition between these divergent programs and goals, and on the other provides a procedure which – despite rivalry and divergence – enables political decisions to be made. This procedure can be circumscribed with the terms "election" and "majority". Accordingly, governments are not formed through consensus and the legitimacy of laws does not depend upon *consensus omnium.* In both cases a parliamentary majority decision is sufficient. Thus democracy is not based on the principle of consensus but on the *principle of majority.* This is not disputed by classical political philosophy. When Locke describes in his second *Treatise* how individuals who are naturally free, equal and independent come together united in a community through agreement, then he is concerned with the origin of the state and civil society as such and the reason behind their legitimacy, and

not with the emergence of concrete political decisions within an already existent state. Locke postulates a consensual agreement solely for the constitution of state and civil society at all; concrete political decisions are made and legitimized not on the basis of consensus but on that of majority. According to Locke, the individuals drawing up the contract are subordinating themselves with the formation of a "political body" to the decision of the majority. To put it pointedly: society-forming consensus puts into force the political principle of majority:

For if *the consent of the majority* shall not in reason, be received, *as the act of the whole*, and conclude every individual; nothing but the consent of every individual can make any thing to be the act of the whole: But such a consent is next impossible ever to be had, if we consider the Infirmities of Health, and Avocations of Business, which in a number, though much less than that of a Common-wealth, will necessarily keep many away from the publick Assembly. ... For where the *majority* cannot conclude the rest, there they cannot act as one Body, and consequently will be immediately dissolved again. ... And thus that, which begins and actually *constitutes any Political Society*, is nothing but the consent of any number of Freemen capable of a majority to unite and incorporate into such a Society. And this is that, and that only, which did, or could give *beginning* to any *lawful Government* in the World. ([13], pp. 350f.)

This solution would hardly be acceptable for morality. It would mean that the institution of morality becomes legitimate through consensus, individual moral norms through voting majorities. The analogy of politics and morality is obviously up against a barrier here. At this point we make a sharp definition between political decisions and measures on the one hand, and moral principles, norms and rights on the other. If the former can be justified using the principle of majority, then this is connected with the fact that political decisions and measures only affect the "external" side of social existence and do not (should not) come into contact with the moral identity of the individual. Neither can the moral autonomy of the individual be annulled through the existence of voting majorities. Yet if the validity of moral norms is only granted through the autonomous self-binding of individuals, then they can only be legitimized by general assent – that is by consensus – and *not* by majorities.

Within a secular, pluralist society, not only will one not be able to identify who has embraced the *true*, concrete view of the good life, but agreement to moral claims by simple pluralities of the individuals involved in a controversy, or by majorities of two-thirds or three-fourths, also will not provide authority, unless *all* can be presumed to have agreed in advance to such procedures. ([5], p. 46)

In this sense autonomy is superior to democracy, and the moral principle of consensus directed *against* the political principle of majority. Theories of consensus in ethics are similar to contractual theories in political philosophy. Yet there is a qualitative difference between these approaches since moral theorists want the validity of moral principles, norms, and rights to be grounded in consensus not majorities. While the practicability of the political principle of majority is beyond question, doubts as to the practicability of the principle of consensus would be well justified. What is there to justify the expectation that in moral debates it will be possible to reach a consensus?

(4) The ethical theories of consensus attempt to solve this problem by reverting to another element of classical political philosophy. In a similar way as that in which the liberal state is understood as being a neutral platform for the regulation of conflicts between divergent individual interests, thus guaranteeing maximal scope for their unfolding, morality should be seen as an institution which protects the individual from outside attacks on his own sphere of legitimate interests and pursuits. The analogy between state and morality is thus based on the fact that a chiefly *negative* function is attributed to both. This implies drastic limitations to the scope of morality. Compared to those views of morality as a regulator guiding the entire spectrum of human behavior (such as utilitarianism, for example), its task is reduced to that of preventing evil. For the problem of consensus this limitation of morality is significant inasmuch as it is imposed in the expectation that the chances of consensus increase when morality is reduced to a "hard core". Bernard Gert, for example, who is not a protagonist of ethical consensus theory in a narrow sense, has expressed his expectation "that if morality is limited to its proper sphere, then one can expect almost complete agreement among rational men on all questions of morality" ([8], p. XV). The basis of this expectation contains two closely connected ideas. On the one hand, the individuals involved in moral disputes are supposed to be acting not "strategically" but oriented towards understanding; that their aim is not to assert their own personal interests, but to find a common solution to the conflict.

The only relevant pressure for agreement comes from the desire to find and agree on principles which no one who had this desire could reasonably reject. According to contractualism, moral argument concerns the possibility of agreement among persons who are all moved by this desire, and moved by it to the same degree ([15], p. 44).

In order to do this they have to disregard their own interests and goals and adopt an objective, impartial "moral point of view". This point of view is defined by the principle of universalizability; that is, by the postulate that moral norms must possess *universal* validity. Turned around, this means that only those norms with a view to general recognition may be accepted as moral. This brings us to the second idea: a view such as this seems only able to exist when the question of a "good life" (which can only be answered in the context of an historically contingent way of life and by the individuals involved in it with their individual and collective preferences) is separated from that of the "morally just" (which requires a generally valid and binding answer). Thus the overall phenomenon of morality must be split into an *evaluative* dimension (a cosmos of values which constitute as a whole an ideal of "good life") and a *normative* dimension (the entire set of rules which is binding for all, i.e. morality in the narrow sense). According to Habermas, the universalizing principle functions "like a knife which cuts between 'the good' and 'the just', between evaluative and strictly normative statements". The field of application for deontological ethics covers "only those questions of a practical nature which may be put rationally and with a view to consensus" ([9], pp. 113f.).

It is obvious where the strength of this approach lies. The idea of an objective "moral point of view", realized in a neutral procedure, allows the existing plurality of particular moral convictions to be recognized without lapsing into a moral relativism. The moral plurality can on the one hand be justified as an expression of freedom. Individuals should have as much freedom to form their own moral ideals as to form their personal and social lives (the epitome of which being moral ideals). Thus there is no real reason to attempt to overcome this plurality and to achieve general consensus in all areas of the good life. Instead of a universal agreement, what we need is a formal framework within which the diverging views concerning the good life may exist alongside one another, and a neutral procedure which allows conflicts to be dealt with in a peaceful manner. Consensus is only necessary on *this* level; and it is also only possible on this level since it does not subject anybody to a particular moral conviction: on the contrary, it guarantees every human being legitimacy and inviolability for his own moral conviction. On the other hand, this magnanimity is not to be had at the price of retreat to a conception of morality, which shies away from strong claims to validity and is content with moral arbitrariness. A concentration on the normative dimension, and the limitations on the scope of morality, is not intended to weaken its authority but to *enforce* it. Thus ethical theories of consensus do not have to be connected with making the idea of consensus absolute. By granting dissent (with regard to the good) just as legitimate a place in the moral life of society as consensus (with regard to the just), they achieve a certain balance between agreement and disagreement.

It is not surprising that, beyond the realm of theoretical ethics, this approach is also attractive for the realm of practical ethics. There are at least two reasons why this also applies to biomedical ethics. Firstly, the situations with which biomedical ethics are concerned are characterized by a strong imbalance between the persons involved. While power, competence, and prestige are usually concentrated on the side of the doctor, the patient is at the mercy of another and extremely vulnerable. This is especially so with regard to technical interventions in human reproduction: the unborn are not even in a position to articulate their own interests. Secondly, there exist between those involved extremely different values. Doctors and patients can have very different views as to which is the correct measure to be taken in a particular situation. Even between the doctors themselves there exist profound differences in the judgment of individual situations or ways of acting in general. With respect to human genetics this has been empirically demonstrated ([16], p. 77). In this problematic situation is a conception of morality which on the one hand allows for the formulation of strict moral obligations, yet on the other leaves enough room for real, existing moral disagreements. The differentiation between a substantial part of morality, which necessarily remains particular and "private", and a universally valid criterion for evaluation, which forms a bridge between the various particular standpoints, provides us with a telling solution to this problematic situation. This differentiation can therefore be used for the field of biomedical ethics in order to constitute beyond all dissent concerning the morally good a sphere of consensus concerning correct procedures. The deontologically hard core of morality is built upon the requirement that informed consent of the person in question be acquired for every case of biomedical treatment. The possibility of free choice thus guaranteed for each individual involved opens up a broad and heterogeneous field for personal preferences and cultural convictions which may be followed with this choice. The neutral procedure of consent is open to every kind of moral "substance" and provides the framework necessary for the coexistence of many different lifestyles and divergent designs for the good life.

III. LIMITS OF CONSENSUS

(5) Last of all, I would like to consider more closely the application of the theory of consensus to specific problems in the field of biomedical ethics. First of all, I will deal on a more general level with the claim that bioethics cleared of the evaluative dimension could provide a "lingua franca of a world concerned with health care, but not possessing a common ethical viewpoint" ([5], p. 5). Secondly, I will specifically go into the problems of technical intervention in human reproduction, and discuss the difficulties which here arise for the principle of autonomy and the required informed consent of the person concerned stemming from it.

Let us begin with a new consideration of the difference between the "good" and the "just". It can hardly be disputed that the differentiation between evaluative preferences on the one hand and a universal decision procedure on the other is meaningful and helpful. It allows particular views of the good life to be removed from the narrow field of generally binding moral norms. At the same time, it should be emphasized that this differentiation is on no account simply given: the morally "just" and the morally "good" are not pre-existent entities or ontologically fixed domains between which there has always been an essential dividing-line since time began. A differentiation of this kind has rather to be *made* within a process of moral reflection. As Habermas appropriately says, it is a case of "abstraction" ([9], p. 116), and this must be achieved again and again. In various cultural and historical circumstances it can lead to various results. Yet if the difference between them can only be ascertained as being an analytical differentiation from a critical point of view, then this may lead – with good reason – to controversy concerning (a) *how* this difference is to be reconstructed in detail and (b) *whether* it is even desirable. This last point is especially relevant when we consider the price to be paid for the separation of the procedure from its contents. The price is renunciation of the identity bringing dimension of morality in favor of a moral universalism which is becoming increasingly indistinguishable from law. Here is one of the roots of the discomfort within the modern world and the resulting fundamentalist reaction. It is the belief that the gain in universality is too expensive at the price of a loss in identity. From this point of view a procedural ethic appears to be the opposite of "neutral". It appears (in the worst case) as renunciation of morality in favor of normative arbitrariness; or (in the best case) as one specific kind of morality. Thus the difference between the good and the just is not a pre-existing, neutral platform, upon which consensus may be formed, but is itself an object requiring consensus.

This non-neutrality becomes obvious if we imagine that there are many different procedures, and then ask ourselves why one procedure has been chosen as binding. We are able to imagine, for example, that matters of moral controversies are decided by drawing lots. The question of whether a certain woman in a certain situation should have an abortion, or whether we should allow gene-technological interventions in the human germ line, would then be answered with the help of a die or a roulette ball. If procedural neutrality were the sole factor, we would have to favor this solution: since what could be more neutral than chance? Of course, there are good reasons for turning down this method in the context of moral decisions. One of these reasons is that it would be impossible to combine a procedure of drawing lots with human moral autonomy. We demand of an acceptable decision procedure and justification criterion that it validate human self-determination. The procedure of informed consent complies with this requirement. Yet by complying with it, it also expresses a certain content which is defined through its function: it serves to guarantee individual self-determination. Thus the procedure is neither purely formal nor neutral. This is confirmed by the arguments underlying the ethics of procedure. It is with good reason that Engelhardt emphasizes that it is only possible to live together peacefully within large, culturally and ideologically heterogeneous societies if the diversity of existing moral convictions is respected. "Respect of the freedom of the individuals" ([5], p. 45) is thus the basis of procedural ethics and must be the starting point for every reflection upon the present moral situation. The right to individual self-determination in all matters of personal lifestyle and the ruling out of an imposition of particular moral points of view are achievements which may only become undone through force – i.e., through the use of immoral means. A society which takes no account of these achievements, and a morality which in the name of "higher" values seeks to disregard them, are thus plainly not acceptable. Yet these are *moral* arguments which contain assumptions which are still to be founded. Engelhardt has recognized and conceded this. He speaks of "circular reasoning": "i.e., reasoning from the notion of ethics as the enterprise of resolving moral controversies without a fundamental resource to force, to the principle of respecting the freedom of the participants in a controversy as the basis of ethics" ([5], p. 46).

The circularity itself of this argument does not seem offensive to me; I consider it to be unavoidable. The question is, however, whether procedural ethics are still at all capable of serving as a "neutral" platform for the achievement of moral consensus, when the question of whether one is prepared to mount such a platform is already a moral one?

(6) The ethics of consensus comes up against a further – within such a framework presumably unsolvable – problem when concerned with technical intervention in human reproduction. Technical interventions in human reproduction concern unborn persons who, due to their not-yet-existence, are obviously not in a position to give their assent to such an intervention. Thus if we are to take the requirement of consent seriously, the entire spectrum of gene and reproduction technology has to be rejected as immoral. There can be no doubt that a "test tube baby" is concerned in a case of *in vitro* fertilization and thus must give its consent to the intervention. Further: not only are all technical interventions in human reproduction immoral; the old-fashioned, natural method of bringing children into the world is in conflict with this principle too. Before now nobody has in advance given consent to his birth and this is not going to change in the future.

As I see it, there are three possible ways of getting around this conclusion. The first demands that when bringing children into the world (be it naturally or technically) we adhere to another moral principle, e.g. the principle of beneficence. This dodge would, however, present us with a serious problem: we would have to assert that it is good for a not yet existent human being to be born. This can of course be disputed, and actually *has been*, and not only by individual, extravagant philosophers. Yet even if we assume that it is good to be born, this would not solve the problem since the theory of consensus ranks the principle of beneficence under that of autonomy: that others may impose upon an individual a particular view of the "good" is precisely what should be avoided. The second possibility would be the readiness to accept belated consent. Thus children who, for example, came into existence via *in vitro* fertilization and/or surrogate motherhood would be asked and – in the case of their assent – the procedure thus made legitimate *ex post*. The question arises then, of course, what should be done if some children did *not* give their consent. Presumably the technology would have to be dismissed. Yet how many would have to refuse to give their consent in order for the procedure to become illegitimate? Apart from the fact that belated consent is unsatisfactory out of principle, it is not always practicable. With the question, for example, of whether the creation of human-animal hybrids may be considered a morally safe option, already discussed in literature, we are faced with the difficulty that such beings would also be belatedly incapable of passing moral judgment on the procedure to which they owe their existence.

Just one path remains. We have to accept that we are unable to obtain the consent of unborn human beings, and yet at the same time adopt the point of view, in order to preserve the principle of autonomy, that we do not *need* this consent. The requirement of consent for all actions is there to protect the autonomy of the individuals involved; future human beings, since they do not yet exist, do not, however, possess an autonomy of this kind. For this reason their autonomy is also incapable of being infringed upon. By "forcing" them to be born through our ("natural" or technical) actions, we far more *create* their autonomy, which is then the prerequisite for their being adopted by the moral community and coming under its protection. This rejoinder is in theory elegant but morally problematic. Its consequence would be that unborn human beings be totally excluded from the realm of morality. They would become morally neutral "material" for the instinct to play and the lust to experiment of already existing humans. Morally valid protests could not even be raised against the most adventurous projects from the horror chambers of the field of eugenics. Precisely those who especially require the protection of morality are incapable of giving their consent and of asserting their interests in the methods underlying morality.

IV. CONCLUSION

If the moral problems raised within the field of gene and reproduction technology are viewed as a *test case* for the efficiency of ethical theories, it is one which comes off badly in the ethics of consensus based on the principle of autonomy. This does not mean that they would thus be "falsified". We know from the philosophy of science that a test is never enough to refute an empirical theory. The negative result of an experiment is usually cause to define the theory more precisely or to restrict its field of validity. The fact that the theory of consensus does not stand up to the problems surrounding human reproductive technologies discussed here is no reason to discard it or to introduce stricter measures to normative theories. Nevertheless, this failure gives rise to certain conclusions regarding both the scope of the ethics of consensus and the problem of justification in the field of ethics in general. I would like to conclude with two of them.

The first conclusion states that the concept of autonomy is a necessary, yet not sufficient basis for the philosophical reconstruction of morality. It is a fact that all autonomy is limited; and consensus cannot be the only criterion for morality. The impossibility of obtaining the informed consent of future persons regarding the manner of their creation should not be disregarded as a problem of peripheral importance or a special case. It is just one – somewhat drastic – example of the limits of moral autonomy. Further: the problem regarding the moral status of the unborn is not limited to the realm of biomedical ethics, but is also a fundamental difficulty within the realm of ecological ethics. If we exclude future beings from the moral community there is no good argument remaining to put a stop to the deterioration of the earth's surface. This means, however, that an adequate ethical system cannot be construed without resorting to other, complementary principles of justification (for example, the utilitarian principle of beneficence). It can only be hinted at here that this result is not only true for the ethical theories of consensus, but may also be generalized. The phenomenon of what is moral is too complex to be reconstrued on the basis of *one* principle. The second conclusion is that differentiations between the good and the just are necessary and helpful but should not be used with the intention of demonstrating the true and never-changing essence of morality. In particular, differentiations of this kind are not a master key for the solution of all moral problems. We are dealing here with ethical differentiations which are not totally removed from the level of moral convictions. To make another analogy to the philosophy of science: it is completely legitimate to differentiate between the "context of discovery" of a scientific theory and its "context of justification" – as long as one does not forget that they are not two separate worlds. In the real-life process of research both levels are connected with each other in such a manner that they are only partially extricable. In the field of ethics the just can, in a similar way, only be analytically separated from the good. The moral life sees the just bound in many ways to the good. This can be made plausible with one simple consideration. If the idea is rejected that differentiations of this nature are ontologically given, then one must presume that they are *made*. This always occurs with a definite goal in mind. Such a goal – and the differentiation serving it – must however themselves be morally justified. The introduction and justification of differentiations such as these is thus always of a circular nature: not necessary in the sense of a *circulus vitiosus* yet presumably so in the sense that we do not have an "Archimedian point" outside of morality from which to achieve consensus.

Department of Philosophy
University of Münster
Germany

NOTES

* Translated into English by Sarah L. Kirkby, B.A. Hons. Exon.

BIBLIOGRAPHY

1. Aristotle: 1949, *Ethica Nicomachea*, The Works of Aristotle, vol. IX, translated into English under the Editorship of W.D. Ross, Oxford University Press, London. Reprint from sheets of the First Edition 1915.
2. Aristotle: 1952, *Politica*, The Works of Aristotle, vol. X, translated into English under the Editorship of W.D. Ross, Clarendon Press, Oxford. Reprint from sheets of the First Edition 1921.
3. Buchanan, J.M. and Tullock, G.: 1962, *The Calculus of Consent. Logical Foundations of Constitutional Democracy*, University of Michigan Press, Ann Arbor, Michigan.
4. Cicero: 1923 (Reprint 1971), *De Divinatione*, trans. W.A. Falconer, Cicero in Twenty-eight Volumes, vol. XX, Loeb Classical Library, Harvard University Press, Cambridge.
5. Engelhardt, H.T., Jr.: 1986, *The Foundations of Bioethics,* Oxford University Press, New York, Oxford.
6. Etzioni, A.: 1968, *The Active Society. A Theory of Societal and Political Processes*, Collier-Macmillan, New York, London.
7. Fichte J.G., 1964: 'Zurückforderung der Denkfreiheit von den Fürsten Europens, die sie bisher unterdrückten. Eine Rede', in R. Lauth and H. Jacob (eds.), *Fichte-Gesamtausgabe*, Bd. I,1 (Werke 1791–1794), Frommann-Holzboog, Stuttgart, Bad Cannstatt, pp. 167–192.

8. Gert, B.: 1973: *The Moral Rules. A New Rational Foundation for Morality*, Harper & Row, New York.
9. Habermas, J.: 1983, 'Diskursethik – Notizen zu einem Begründungsprogramm', in J. Habermas, *Moralbewußtsein und kommunikatives Handeln*, Suhrkamp, Frankfurt/M., pp. 53–125.
10. Hobbes, T.: 1839 (Second Reprint 1966), *Leviathan*, The English Works of Thomas Hobbes of Malmesbury, vol. III, Ch. 17, pp. 153–159, Scientia Verlag, Aalen.
11. Kant, I.: 1964, 'Zum ewigen Frieden. Ein philosophischer Entwurf', in W. Weischedel (ed.), *Werke*, vol. 6, Insel Verlag, Frankfurt/M., pp. 191–251.
12. Locke, J.: 1959, *An Essay Concerning Human Understanding*. Collated and Annotated, with Prolegomena, Biographical, Critical, and Historical by Alexander Campbell Fraser, New York, Dover Publications.
13. Locke, J.: 1970, *Two Treatises of Government. A Critical Edition with an Introduction and Apparatus Criticus by Peter Laslett*, 2nd ed., Cambridge University Press, Cambridge.
14. Rawls, J.: 1971, *A Theory of Justice*, Harvard University Press, Cambridge.
15. Scanlon, T.M.: 1986, 'A Contractualist Alternative', in J.P. DeMarco and R.M. Fox (eds.), *New Directions in Ethics. The Challenge of Applied Ethics*, Routledge and Kegan Paul, New York & London, pp. 42–57.
16. Wertz, D.C. and Fletcher, J.C. (eds.): 1989, *Ethics and Human Genetics. A Cross-Cultural Perspective*, Springer, Berlin, Heidelberg, New York.

LUDGER HONNEFELDER

CONSENSUS FORMATION FOR BIOETHICAL PROBLEMS

Comments on K. Bayertz's Paper on "The Concept of Moral Consensus. Philosophical Reflections"

For the justification of actions that a multitude of agents is responsible for and/or affect a multitude of others a consensus about moral judgements is necessary. This consensus can be subject to dispute on a number of levels. There may be dissent in the judgement about an individual action; or the dissent may be about the formation of norms which are the basis for the justification of a class of actions; or there may be dissent about the principles on the middle or higher levels of universality which determine the formation of such norms; lastly, the dissent may concern the entire metaphysical, religious or other interpretations of the meaning of life from which the obligatory force of those principles may be derived. If the agents are all members of a group that share a comprehensive yet specific ethos, i.e., a determined conception of the good life, dissent will only concern particular actions or perhaps the formation of norms but not the acceptance of their underlying principles. The necessary formation of a consensus presents itself as a problem of casuistry or of the finding of specific norms. The matter differs for communities such as modern society which allow room for the development of more than one specific ethos. For here the dissent may be not just about specific norms but also about the principles which determine them. The problem is not just which common moral principle can be appealed to if a consensus is to be found, but also which strategy is to be chosen to achieve the desired result. While a group with a common ethos will aim at a consensus as complete as possible, this is not a reasonable goal for a pluralistic society. If consensus within a specific ethos is desirable, a society which does not share such an ethos must admit dissent, i.e., freedom. Here consensus may only be aimed at where it is necessary, i.e., where many agents of different moral convictions act together and/or the consequences of such actions are relevant to many or all.

Kurt Bayertz (ed.), The Concept of Moral Consensus, 59–63.

As regards consensus formation in a modern pluralistic society there can be no doubt that one has to take account of the fact that the defense of actions and norms has become an explicit part of the modern ethos itself and demands a corresponding way of forming moral judgements. But this does not mean, as Bayertz's genealogy suggests, that consensus did not play a part in any pre-modern ethos. Consensus formation for ethical problems has not become necessary, because moral patterns enforced by an authority have historically been superseded by those which have to be agreed upon in an open discourse, but because moral patterns shared within a relatively unified and closed ethos have been replaced by a multiplicity of ethical models which share only a minimum of premises. Moral patterns can be enforced by external repression only for a limited time. They have institutional authority (which may be limiting or even repressive for some agents) because they are consensually shared, not because they are enforced by an external authority, though the consensus may to a large degree remain implicit. Ethics based on an ethos does not contradict moral autonomy, nor does an ultimate interpretation of moral obligation which refers to a conception of the meaning of life anchored in transcendency, as long as the demands of the ethos and the truth of the interpretation can be affirmed rationally by the agents. The history of the concept of conscience shows that it is precisely its Christian interpretation relating it to God which opens the way for the discovery of the fact that moral judgements have a two-tiered and reflexive nature, and thus to Kant's concept of autonomy (cf. [3]).

Modern differentiation and pluralization lead to a decrease in the number of moral patterns which are implicitly shared and consequently to an increased demand for explicit consensus formation in the area where norms for acting together have to be found. I agree with Bayertz against MacIntyre that this process must not be interpreted as negative [1]; nor am I so sceptical about the possibilities of resolving it as are Jonsen and Toulmin [6]. Only if, as MacIntyre does [7], you chose a situation as your standard in which society, culture, religion and ethos form a far reaching unity will the development of a pluralistic society appear to be a degeneration. And only if you do not distinguish appropriately between the independent plausibility of the moral principles and the religious or other interpretations of the meaning of life which may be used to justify their plausibility, will the chance to gain universal plausibility for the moral principles appear to be vanishing alongside the pluralization of the background premises. Of course, things have changed dramatically. More room for the autonomous subject is equivalent to an increased burden in the formation of individual and collective identities; the disappearance of areas of given moral consensus makes the subsequent task of forming a consensus a more difficult problem.

The problem of consensus formation is accentuated by an increasing number of options for action which is a consequence of the technical application of modern science, especially in the biomedical field. On the one hand, the consequences and side effects of the new options for action are so far reaching that you cannot leave them unregulated. On the other hand, the moral principles, even if we suppose that consensus about them can be reached, are not sufficient to extract from them limits that can be agreed upon. Science only describes a framework of conditions for our actions, but it does not in itself give them any orientation. Yet, because our options for action are of a completely new kind, we cannot rely on our experiences concerning previous regulations.

It has to be expected, therefore, that consensus formation for bioethical problems is a very complex process with narrowly defined goals. As with any moral discourse, so in the formation of consensus for bioethical problems you have to distinguish between the area for which consensus may be assumed and the area for which a consensus has yet to be found. As regards this distinction it makes sense to interpret the moral judgement as a two-tier process in which the actions and norms are examined in comparison to already accepted principles. Amongst these principles is one which may be called the *highest practical principle* or the most general supposition for any ethical discourse and which plays a special, though at once limited, role. It determines the form and obligatory force of any moral judgement and any formation of a moral consensus; and it has a plausibility which may (for an ultimate justification: has to) be subject to a deeper justification through a comprehensive metaphysical, religious or other interpretation, but which commands authority independently of such interpretations. This highest principle – in any possible phrasing – will yield no further content than the form which it determines. Its demand, only to do what is regarded as morally obligatory, does not specify any content over and above the description of the moral subject as the decisive standard, and thus as far as consensus formation is concerned only yields the demand that nobody may be forced to act against his or her conviction.

If the aim is to achieve a consensus about norms, then a consensus about the most general and formal principles of judging and acting morally does not suffice. We need *"middle principles"* which describe generally acceptable material standards. Identifying these principles is the most important problem for consensus formation, especially in the field of bioethics. For here we are concerned with the determination of those needs and interests which are considered to be basic in as much as they form a standard for what is indispensable or unacceptable. If the "middle principles" are not supported in advance by a particular concept of the good life, i.e., a particular normative picture of mankind, the search for consensus is bound to fail, since there will be many interpretations of the criteria for what it is to lead a good life. Because nature as considered by science does not yield such criteria, success of consensus formation for bioethical problems depends on the availability of a *concept of nature* which includes a practical dimension and thus provides the necessary orientation. We find the nucleus of such a concept of nature in the human rights which define certain interests and needs as indispensable (and consequently as subject to protection) with respect to human nature. Human rights are accepted world-wide independent of their socio-cultural origin, which shows that it is not without prospects to anchor the consensus formation in "middle principles" which secure the conditions of the possibility of human life in nature (cf. [4]).

However, such "middle principles" are not sufficient for the formation of a consensus on moral norms. For the "middle principles" only state framework conditions by naming fundamental goods which neither the individual nor a community can achieve all at once without conflict. A successful consensus formation for norms must thus include rules and criteria for the *evaluation of goods*. But the problem of the evaluation of goods refers – over and above the natural and fundamental needs and interests – to a specific *ethos* which expresses a concrete concept of what is considered to be a good life (cf. [5]). In the field of bioethics this becomes apparent with respect to the concepts of health and disease. On top of natural elements they include normative ones which are not independent of what in any given society is considered to be a good life, even though they only define the margins of the good life. Consensus formation is therefore ultimately also a question of how much of a specific ethos is considered to be common ground by those concerned. This in turn depends on whether the ethos in question can be apprehended and acquired as a structured and sensible whole.

University of Bonn
Germany

BIBLIOGRAPHY

1. Bayertz, K.: 1994, 'Introduction: Moral Consensus as a Social and Philosophical Problem', in this volume, pp. 1–15.
2. Bayertz, K.: 1994, 'The Concept of Moral Consensus. Philosophical Reflections', in this volume, pp. 41–57.
3. Honnefelder, L.: 1982, 'Praktische Vernunft und Gewissen', in A. Hertz *et al.* (eds.), *Handbuch der christlichen Ethik*, vol. 3, Herder, Freiburg, Basel, pp. 19–43.
4. Honnefelder, L.: 1991, 'Person und Menschenwürde. Zum Verhältnis von Metaphysik und Ethik bei der Begründung sittlicher Werte', in W. Pöldinger and W. Wagner (eds.), *Ethik in der Psychiatrie*, Springer, Berlin, Heidelberg, New York, pp. 22–39.
5. Honnefelder, L.: 1991, 'Güterabwägung und Folgenabschätzung in der Ethik', in H.-M. Sass and H. Viefhues (eds.), *Güterabwägung in der Medizin*, Springer, Berlin, Heidelberg, New York, pp. 44–61.
6. Jonsen, A.R., Toulmin, S.: 1988, *The Abuse of Casuistry. A History of Moral Reasoning*, University of California Press, Berkeley, Calif.
7. MacIntyre, A.: 1981, *After Virtue*, University of Notre Dame Press, Notre Dame, Ind.

HENK A.M.J. TEN HAVE

CONSENSUS, PLURALISM AND PROCEDURAL ETHICS

Responding to the interesting contributions of Engelhardt and Bayertz, I will concentrate on what I consider the two main issues articulated by both authors, *viz.* (1) the diagnosis of post-modern society as radically and fundamentally pluralistic, and (2) procedural ethics as the therapy for the problems of pluralism and as the privileged solution to moral controversies in large-scale states.

I. INTERPRETATIONS OF PLURALISM

Engelhardt's well-known thesis is that we should distinguish between the secularized, pluralistic society on the one hand and the many particular moral communities on the other hand ([3]; [4]). This distinction not merely implies a difference of argumentative level or scale, but it flows, first of all, from a specific evaluation of the human predicament. In small communities, people share basic moral values; in pluralistic society, individuals are moral strangers. Consensus, considered as unanimity of opinion, can only be accomplished on the level of the particular moral communities, – and necessarily so, for two reasons:

a) Because at the public, social level consensus would imply the tyranny of a majority view; in post-modern societies it is morally imperative to respect competing visions of the good life.
b) Unanimity of opinion with regard to moral matters is necessarily bound to a specific moral community.

Bayertz introduces a similar distinction, *viz.* between "the good" and "the just". Morality has an evaluative dimension ("a cosmos of values which constitute ... an ideal of 'good life' ") and a normative dimension ("the entire set of rules which is binding for all") ([1], p. 49).

Kurt Bayertz (ed.), The Concept of Moral Consensus, 65–74.

Having made these distinctions, both Engelhardt and Bayertz face at least some ambiguous consequences. Since there are competing visions of the good life, the state should not reward or penalize particular conceptions of the good life; it must be neutral in matters of morality, providing a framework within which various moral conceptions can be pursued. For Engelhardt the basic difference seems to be between morality and politics. On the moral level, individuals are embedded in communities; they can only flourish because they belong to a community sharing a vision of the good life. Ethically, this level is the most interesting since it generates normative meaningful accounts of moral controversies such as reproduction, relationship and sex. Here consensus is certainly possible; in part it is even an a priori condition for being a member of the particular community. On the political level controversy and dissent rather than consensus are likely; consensus would mean the unauthorized domination by some "moral majority". The state, however, must leave people free to live as they think best. In order to resolve controversies peaceably, some minimum agreement on the conditions of cooperation is necessary, and Engelhardt argues that mutual respect and consent can be regarded as such basic conditions ([4], p. 32).

Bayertz also locates the problem of consensus within the historical and philosophical context of the polarity between state and individual. But for him the fundamental difference is not between the moral and political level, but between substance and form. Contrary to Engelhardt, he argues that consensus is only possible on the level of a formal framework within which the diverging views concerning the good life may co-exist ([1], p. 49). Such consensus, however, is shallow; it is devoid of almost any interesting moral dimension, – but not completely, as Bayertz shows: procedural neutrality is combined with a certain moral content ([1], p. 52).

The basic motivation to introduce the above-mentioned distinctions in the bioethical debate concerning reproductive technologies is the moral pluralism of contemporary society. It would be rather foolish to deny the very existence of pluralism but it is nonetheless unclear how to interpret it and to clarify its moral significance. Engelhardt's characterization of the pluralistic condition of post-modern society seems too strong. His interpretation of the secularization of the western world is guided by general distinctions and commonplace categories: rootless secular cosmopolitans are contrasted with orthodox religious believers, and both are involved in irresolvable controversies ([4], p. 17); for those living within religious communities everything has transcendent significance whereas in the secular, international society the world is without transcendent significance ([4], pp. 27f.); and Christianity "has become a collection of sects and cults" ([4], p. 25). Data from empirical studies (especially in the sociology of religion) suggest that interpretation of contemporary societies as fundamentally secularized should be nuanced and differentiated ([6]; [7]; [8]; [11]). In at least three respects, the pluralistic condition of post-modern societies is more complicated than it may look from a bird's eye view.

(1) Traditional religion has declined; but from that historical process it may not be concluded that there has been a similar decline of transcendent religiosity. On the contrary, the European Value Systems Study Group survey data, based on research in 10 West-European countries ([6]; [7]), show that for the overwhelming majority of the population of these countries there is no loss of transcendent meaning of life and world, – there has been an enormous change in the institutional setting, in the traditional modes of religion, but not so much a decline of religiosity or a decay of transcendent value systems. In most countries traditional religion is no longer a pervasive and powerful factor, but it does not follow from this observation that it has lost its cultural and political influence in societies. In a relatively homogeneous area as Western Europe, societies also differ in various respects from one another. From his research data, Halman concludes that sociological theories of social change and modernization are usually formulated at a too general and abstract level, conceptualizing modern society as if there exists indeed just one modern society differing in all aspects from premodern society ([7], pp. 370–374). Even within post-modern societies, heterogeneity and nation-specific differences prevail, that are anomalies and inconsistencies from the perspective of modernization theories. In the European Value Systems survey it is a remarkable finding that the Netherlands is one of the most secularized European countries: more than 50 % of the population is not a member of any church (compared to an average 8 % in other European countries) ([12], pp. 24ff.). But low church participation is not coincident with a decline in general religiosity, while a high level of church participation is not necessarily associated with traditional faith or orthodoxy. The United States, on the other hand, seems to be the most modernized country, whereas value study data show (much) higher levels of religious feelings, orthodoxy, church participation as

well as confidence in the church than in Northern European countries ([7], pp. 51ff.)

What such findings indicate is the discrepancy in post-modern societies between the endorsement and the justification of values. As Engelhardt cogently argues, human reproductive technologies are quite differently valued depending on different interpretive frameworks; values are justified from a specific normative understanding of human nature and from a transcendental or secular perspective on life and world ([4], p. 23). But although the justificatory strategies differ, the values itself may not be incongruous. Among the majority of the population of the examined countries there seems to be no radical change. For example, in marital values, most people share traditional views on sexuality even in those countries (e.g., Denmark) where sexual freedom is widespread. What has changed is the increased tolerance for previously not accepted behavior, but not an increase in an active involvement in such behavior ([7], pp. 183ff.).

(2) For the sake of argument, Engelhardt has contrasted two parties in the struggle for consensus in large-scale, democratic states: rootless secular cosmopolitans and orthodox religious believers. However, in real life, both parties obviously are a rare species. Value research indicates that on average two-thirds of the respondents in Western European and North American countries identify themselves as religious believers ([7], pp. 59ff.). The percentage of respondents considering themselves as convinced atheists is very small (10% in France, 4% in Great Britain, Denmark, The Netherlands, and 1% in the U.S.A.). On the other hand, the percentage of respondents completely in agreement with all doctrines of Christianity is also comparatively low (an exceptional 50% in the U.S.A. and Ireland, but much lower in Western European countries: 5–10%). For example, the percentage of Dutch respondents agreeing with the orthodoxy of the Christian churches is seven. Even from those who go weekly to church only 18% fully agree with the official doctrines ([12], p. 19). However, most people subscribe to some elements or subsets of the doctrines.

Value research makes clear that many people in post-modern societies no longer have religious and moral values that are homogeneous and consistent from a doctrinal or theoretical perspective. In fact, almost all values emphasized by religious doctrines and their moral systems, are endorsed by the majority of Western populations, but the justification of the values is no longer derived from those doctrines and systems. In practical life, religious and moral world views are fragmented; they are the result of *bricolage*, – and therefore less coherent than the traditional view of institutionalized religion. It means that the rationale for what is valued in life is not derived from a coherent and closed system but the result of an eclectic and pragmatic activity of moral figuration, "processes of inclusion, exclusion, and reconfiguration at work in creative moral thought" ([13], p. 76). From the coherence of a system or tradition, however, its seems as if individuals have a contingent, sometimes even inconsistent, mosaic set or collection of religious and moral values.

These observations show that an important distinction should be made: on the theoretical level, explicit construction of consensus is a perplexing moral problem, particularly in an Engelhardtean world of antagonistic orthodox believers and secular cosmopolitans; on the practical level of the lifeworld, however, there is *de facto* a lot of agreement and overlapping consensus concerning moral values.

(3) The last point may be further elaborated by referring to some interesting statements in Engelhardt's chapter: Secular cosmopolitans are those "who live outside of religious or other particular normative traditions..." ([4], p. 28); they live in "the secular, international society...that spans from Brazilia to Montreal ... from Buenos Aires to Paris..." ([4], p. 28). Although it is unclear whether the author means living outside of *any particular* normative tradition or living outside of *any* normative tradition, the question in both cases is whether that is really possible. This is one of the issues in the debate between libertarians and communitarians. A society seems to be bound together by a complex of norms and values adhered to by the majority of its members. The important role of such binding norms and values is accentuated by the concept of *"civil religion"*: basic values that symbolize the uniqueness and legitimation of a society and that endow its members with an identity.

These basic values can not be changed without disturbing the social ordering of a society; they are constitutive rules rather than regulative rules. What members of societies have in common in sharing value orientations is more important than the points of difference among them. Recent research made clear that in most Western European countries there are such common, constitutive values: freedom, equality, and solidarity ([2]; [6]). From time to time, these values are quasi-religiously celebrated on special occasions to re-emphasize national identity (for example, Memorial Day). It would be political "suicide" for a politician to seriously question one of these basic values. In the Netherlands, for instance, even in times of scarcity nobody will question the principle of equal access to health care or the principle of a proportionate distribution of financial burdens according to income. Recent data show around 70% of the population subscribing to the view that higher income-groups should be made to pay higher health insurance premiums than lower income-groups, a view which also turned out to be endorsed by two thirds of all respondents from higher income-groups [9].

Basic societal values are abstract: they allow for different interpretations. But that is their power: everybody can endow these values with his own interpretation instead of promoting his interpretation in contrast to these values. These empirical investigations show therefore that there is a high level of consensus with regard to basic values, but dissensus with regard to the interpretation of these values.

II. WHAT KIND OF ETHICS?

Bayertz and Engelhardt address the issue of consensus with such transparency that the specific conception of ethics involved in their exposition manifests itself immediately. Particularly, their suggestions to settle the philosophical problem of how to resolve rationally controversies on the state level raise questions about the nature of ethical understanding. I will briefly discuss three issues coming to mind when reading the above chapters: (1) the procedural concept of ethics; (2) the moral significance of mutual respect and tolerance; (3) the neutrality of the common moral language.

(1) Engelhardt's conclusion is that despite the pluralistic character of post-modern societies we should (and in fact we can) develop strategies for speaking across gulfs of moral discourse. There is in fact a moral grammar, a common neutral language to prevent moral war and to guarantee a peaceable society. In Engelhardt's distinction of the two levels of moral discourse, the most interesting task of ethics is on the second level: promoting and defending the general secular moral language of mutual respect. That is an important task but it seems to flow from a rather thin conception of ethics, *viz.* ethics as regulation of social relations through peaceable negotation. This is a formal concept of ethics since more substantive moral issues are addressed on the lower level of particular communities. The abstraction from substantive issues

makes this concept attractive, creating opportunities for a peaceable resolution of controversies. But precisely the theoretical characteristics of such a formal concept of ethics is nowadays under close scrutiny of moral philosophers.

Bernard Williams, for example, has criticized modern ethics as a reductivist enterprise. Ethics, in his view, does not respect the concrete moral subject with its personal identity. It requires that the subject gives up his first personal point of view and exchanges that for the universal and impartial point of view of anyone. That is an absurd requirement because the moral subject is requested to give up what is constitutive for his or her personal identity: "...it is to alienate him in a real sense from his actions and the source of his action in his own convictions" ([15], p. 49). If ethics is about how we should live, we cannot abstract from our own subjectivity. Similar issues are discussed in Charles Taylor's recent work [14]. One of Taylor's questions is whether ethical language and practices are thinkable without a real community, a genuine social identity? In his view, morality and identity are two sides of the same coin. To know who we are is to know the moral sources to which we appeal. The community, the particular social space to which we belong, is the centre of our ethical experience. The use of ethical language depends on a shared form of life. This is in fact the old Wittgensteinian idea that our understanding of language is a matter of picking up practices, being inducted into a form of life. We have our moral sources, even in scientism, post-modern deconstruction, and in Engelhardt's common neutral language. There are moral sources beyond ourselves which explain our commitments to non-violence, justice, or peaceable resolution.

(2) The emphasis on mutual respect and tolerance is in accordance with the high level of appreciation of these values in western societies. But so are other values. Equality, for example, is also highly valued, and in some countries even more than individual freedom ([7], p. 314). It is therefore not evident why we should prefer the language of mutual respect and tolerance to speak "across gulfs of moral discourse" ([4], p. 20). But, more importantly, it can be doubted whether this preference is really helpful. Does it offer a solution to the problem of moral pluralism? It is Engelhardt's thesis that substantive moral arguments are relevant within the moral community to which one belongs. Disagreements concerning substantive moral questions are always bound to particular moral communities. Since this is the case, and in order to avoid confrontation, a polemic *between* moral communities must

therefore be avoided. Such a confrontation leads to nothing, or in the worst cases to moral warfare, and finally to tyranny. Engelhardt's solution, introducing a second-order level establishing a procedural moral framework, is in fact an attempt to escape from our pluralistic situation, to escape from the fundamental moral differences in modern society. The only way to discuss moral issues in our society is to speak the language of mutual respect, – all other moral languages must be pacified. So in the end it seems that the problem of pluralism is not so much resolved but rather covered up. The really perplexing controversies, which are substantive, are avoided and in fact relegated to a lower level of philosophical significance and interest.

(3) But why should we abstain from our particular moral language in favor of a neutral common language? This question points to an important problem: how neutral is the common neutral language? Is this language itself not the specific moral language of a specific moral community? Is this language itself not the expression of a commitment to a certain "hypergood", in particular the demands of universal and equal respect and of self-determining freedom – primal values in the liberal tradition? Bayertz argues on the one hand that the practice of free and informed consent is a neutral procedure "open to every kind of moral substance..." ([1], p. 50), on the other hand that it is also the expression of a certain content, *viz.* the moral primacy of individual self-determination ([1], p. 52). He agrees that basically the procedure cannot be separated from its contents: it is "neither purely formal nor neutral" ([1], p. 52). The values of mutual respect and tolerance as well as rights to privacy are not decontextualized standards but themselves expressions of community-bound agreement.

The non-neutrality of the "neutral common language" has recently been underlined by sociological studies, showing how liberalism and individualism are value orientations very much characteristic of American bioethics [5]. Liberalism is not neutral about what would count as a good society and what would be good for individuals. Moreover, autonomous choices are only possible in a shared cultural structure that provides individuals with meaningful options. The moral language of mutual respect and tolerance can therefore not be neutral regarding the conditions that are essential to its survival; it should guarantee "...the existence of a pluralistic culture which provides people with the range of options necessary for meaningful individual choice" ([10], p.893). It seems that Engelhardt's procedural conception of ethics is not only itself community-bound, representing the language of the community of liberal and analytic bioethicists, but is also self-defeating as long as it is neutral towards the very socio-cultural conditions in which it is a worthwhile approach.

But if that observation is true, the following conclusion seems more or less inescapable: his resolution of the problem of pluralism is nothing less than a philosophical *coup d'état*. The neutral moral language as a *de facto specific* moral language is pretending to be *the de jure* moral language for all of us interested in resolving issues peaceably. The problem of consensus is removed from the philosophical agenda through a discursive, rhetorical strategy, putting in charge one particular moral language at the expense of all other moral languages prevailing in our pluralistic societies.

Department of Ethics, Philosophy and History of Medicine
Catholic University of Nijmegen
Nijmegen, The Netherlands

BIBLIOGRAPHY

1. Bayertz, K.: 1994, 'The Concept of Moral Consensus. Philosophical Reflections', in this volume, pp. 41–57.
2. Borg, M.B. ter: 1990, 'Publieke religie in Nederland', in [11], pp. 165–184.
3. Engelhardt, H.T., Jr.: 1986, *The Foundations of Bioethics*, Oxford University Press, New York.
4. Engelhardt, H.T.,Jr.: 1994, 'Consensus: How Much Can We Hope For? A Conceptual Exploration Illustrated by Recent Debates Regarding the Use of Human Reproductive Technologies', in this volume, pp. 19–40.
5. Fox, R.C.: 1989, 'The Sociology of Bioethics', in R.C. Fox, *The Sociology of Medicine: A Participant Observer's View*, Prentice Hall, Englewood Cliffs, New Jersey, pp. 224–276.
6. Halman, L.*et al.*: 1987, *Traditie, secularisatie en individualisering. Een studie naar de waarden van de Nederlanders in een Europese context*, Tilburg University Press, Tilburg.
7. Halman, L.: 1991, *Waarden in de Westerse wereld. Een internationale exploratie van de waarden in de westerse samenleving*, Tilburg University Press, Tilburg.
8. Inglehardt, R.: 1990, *Culture Shift in Advanced Industrial Society*, Princeton University Press, Princeton, New Jersey.
9. Janssen, R. *et al.*: 1987, 'Solidariteit en het ziektekostenverzekeringsstelsel', *Gezondheid & Samenleving* 8, 2–9.
10. Kymlicka, W.: 1989, 'Liberal Individualism and Liberal Neutrality', *Ethics* 99, 883–905.

11. Schreuder, O. and Snippenburg, L.van (eds.): 1990, *Religie in de Nederlandse samenleving. De vergeten factor*, Amboboeken, Baarn.
12. Schreuder, O.: 1990, 'De religieuze traditie in de jaren tachtig', in [11], pp. 17–41.
13. Stout, J.: 1988, *Ethics after Babel. The Language of Morals and Their Discontents*, Beacon Press, Boston.
14. Taylor, C.: 1989, *Sources of the Self: The Making of the Modern Identity*, Cambridge University Press, Cambridge.
15. Williams, B.: 1988, 'Consequentialism and Integrity', in S. Scheffler (ed.), *Consequentialism and its Critics*, Oxford University Press, New York, pp. 20–50.

HELGA KUHSE

NEW REPRODUCTIVE TECHNOLOGIES: ETHICAL CONFLICT AND THE PROBLEM OF CONSENSUS*

I.

In the field of new reproductive technologies ethical controversy has followed close on the heels of scientific discovery. When Robert Edwards and Patrick Steptoe published, in *Nature* in 1969, the first account of the fertilization of a human egg outside the body, the Archbishop of Liverpool immediately condemned the experiments as 'morally wrong' and Baroness Summerskill, the social reformer, supported them as a morally uncontroversial mode for overcoming infertility ([9], p. 88). The ethical debate on new reproductive technologies has continued ever since and shows little signs of abating. This raises questions not only about the morality of particular reproductive technologies themselves but also about the nature of ethics and about the central philosophical and practical role recently attributed to the idea of consensus.

At first glance, it may seem somewhat odd that the idea of consensus should get hold of the philosophical imagination at a time characterized by unprecedented ethical conflict. Not only are proponents of various reproductive technologies engaged in shrill and apparently interminable battles with those who oppose them, but the cacophony of voices extends to ethical theory itself: utilitarian conceptions of the good are matched against theories of individual rights, of human virtues and Kantian notions of absolute moral laws. Nonetheless, it is precisely in times of apparently irresolvable moral conflict that the idea of consensus is likely to become the focus of attention. On the level of ethical theory, it will become attractive to search for principles or norms to which everyone could agree under certain conditions; and, on the level of practice, it will become necessary to look for agreement on an ethical framework or

Kurt Bayertz (ed.), The Concept of Moral Consensus, 75–96.

procedure which would allow one to resolve moral conflict in peaceful ways.

I shall begin by locating consensus theories in the field of ethics and show that they are faced with a number of difficult problems. While these ethical theories are advanced as neutral frameworks for the resolution of moral disputes, those who have attempted to derive a substantive account of morality or content-full procedure for the resolution of moral conflict from them can quite properly be accused of having smuggled their own substantive account of ethics into the supposedly neutral framework. This means that these theories will not be able to help us resolve the most vexing practical problems raised in areas such as embryo experimentation, surrogacy, and the like. Nonetheless, it seems that a procedural framework – based on mutual respect and rational discourse – can serve as the vehicle for the resolution of moral disputes. After distinguishing between different "levels" of ethical conflict and consensus, I shall suggest that properly constituted and consensus-oriented ethics committees are the most appropriate vehicle for resolving ethical conflict. Such committees will do this not by presenting us with "the truth", but rather by presenting society with rational arguments and, hopefully, morally acceptable solutions.

II.

"Consensus Development Conferences" and "Consensus Statements" on morally contentious issues in medicine and the biomedical sciences bear witness to the fact that consensus is highly valued. There is, of course, a straightforward reason why this should be so. Without broad consensus on a morally contentious issue or proposal, it is unlikely that the proposal will find acceptance in the public arena. In addition to that, consensus, in the sense of general moral agreement, may, in the famous words of Lord Devlin, be seen as the cement which binds society together. Without such a shared moral view, he thought, society would disintegrate [8]. While one may want to disagree with Devlin's judgment regarding the fragility of society, or on the measures he thought necessary for enforcing a shared moral view, it would nonetheless seem true that broad agreement on fundamental values and beliefs will contribute to the smooth functioning of society. A society with a shared moral view is less likely to be torn by internal strife, and its members will more readily cooperate with each other towards shared ends and in accordance with agreed principles or norms. Hence, if one values harmonious social relations and absence of strife, one will have a further reason for regarding consensus as desirable.

The fact that moral consensus will generally facilitate decision-making, or contribute to the smooth running of societies, does, however, not entail that consensus is valuable in itself. Not only does it seem quite plausible to hold that there is value in rich moral diversity and in the intellectual and practical challenges to which it gives rise, but there is also the question of whether we have any grounds for believing that there is a necessary connection between consensus and moral truth. While Aristotle thought "that that which every one thinks really is so" (*Ethica Nicomachea*, X 2,1173a, [3]), contemporary philosophers have largely followed Plato and Kant, who regarded consensus on a moral matter as a merely contingent historical fact to which no moral authority or truth claims could be attached. Indeed, if one recalls that Aristotle's pronouncement that a slave is but "a living possession" and an "instrument" (*The Politics*, I 4, 31–32, [4], p. 1131) but restated the common view of the time, then one would immediately want to banish the idea that there is any connection between consensus on a moral matter and moral truth.

The belief that consensus in the sense of a shared moral belief is connected to moral truth would seem to commit one to moral relativism, and to the view that opinion polls could determine the correctness of moral judgments. This would mean that disputes such as those between the Archbishop of Liverpool and Baroness Summerskill, to which I referred at the beginning of this paper, could be settled rather simply. If a sufficiently large majority of the British population had indicated that they agreed with, say, the Baroness, then the Archbishop's pronouncement that the experiments are morally wrong could be dismissed as a simple factual error. This is enough, it seems to me, to show that consensus, understood as "majority view" or "shared moral belief", is not an adequate basis for the grounding of moral claims.

Nor is it this understanding of "consensus" as common or majority view that modern proponents of ethical theories of consensus have in mind. Rather, as Kurt Bayertz has pointed out [5], in these ethical theories the concept of consensus is linked to "autonomy" and "consent", that is, to the voluntary and free acceptance of certain ethical norms by morally autonomous individuals. In other words, consensus plays a *legitimizing* role for the acceptance of norms, in much the same way as it has traditionally done in political philosophy and democratic theory. For political philosophers the question was: how can the state legitimately exercise power over free and independent individuals? The still widely accepted answer was that the exercise of state power is legitimate if all free and independent individuals have given their consent. In other words, consensus – now understood as universal assent – is seen as the necessary condition for the legitimacy of the democratic state.

In a similar vein, Bayertz explains, proponents of ethical theories of consensus are seeking to ground ethical principles and norms in universal assent. While individuals are seen as morally autonomous, that is, as the makers of their own morality, communal living entails the acceptance of some binding norms. To have moral authority, these norms require universal assent; in other words, consensus is seen as the necessary condition for the legitimacy of ethical norms, just as it is for the legitimacy of the state [5].

"Consensus", understood as universal assent, is central to a number of quite different theories – ranging from the contemporary contractualist theories of John Rawls [24] and his followers, over the bioethical framework put forward by H. Tristram Engelhardt, Jr. ([10]; [11]), to the discourse ethics developed by Jürgen Habermas and others ([13]; [2]). While there are important differences between the various ethical theories of consensus, we will not so much focus on these differences as on the theories' common features. As we have already noted, all regard universal assent as a necessary condition for the legitimacy or authority of ethical norms; and a number of them take as their starting point not the real, empirical consent under actual circumstances of all those affected, but rather the hypothetical consent of idealized people who are subject to certain conditions.

To begin with, I want to look at some direct practical implications of the theoretical device of idealized or fictitious consent. This discussion will then lead us on to another feature of consensus theories of ethics – the division of the ethical into an evaluative/private and a normative/public realm.

III.

When people find themselves in a Rawlsian "original position" ([24], pp. 11–22), or a Habermasian "ideal speech situation" ([13], pp. 107f.), they are stripped of all their present interests, motivations, life experiences, and the like. For Rawls, the agents who are party to the original bargaining session are presumed to be ignorant of their actual place in society, but are presumed to be rational and self-interested. For Habermas, the agents must be motivated by no desire, other than the desire to jointly find and agree on ethical principles or procedures. Agreement on normative principles or procedures is possible precisely *because* the heuristic devices employed ensure that the agents are not motivated by their present desires, interests, or goals, but are adopting a "universal" or "impartial" point of view.

This means that it is likely that there will always be a gap between the hypothetical consent elicited under "ideal" conditions, and the "real" consent of actual people, who are shaped by particular life experiences, who have certain interests, motivations and visions of the "good life". In other words, while it might be correct that the Archbishop of Liverpool and Baroness Summerskill would, under some idealized or fictitious conditions, agree on the morality of embryo experimentation, in the real world, here and now, they are disagreeing on the matter. Can this disagreement be overcome? If not, what does this entail for consensus theories of ethics?

H. Tristram Engelhardt, Jr., has argued, convincingly in my view, that the West (where most of the debates regarding the propriety of new reproductive technologies are taking place) is characterized by a multitude of visions of "the good life" [10]. These visions, deeply embedded in cultural and religious traditions, are the back-drop against which the morality of various actions is judged. While sex, for a "Yuppie", unencumbered by deep cultural or religious beliefs, may be nothing more than good recreational fun, it will have quite a different meaning for someone who, for example, lives in a traditional religious community, where sex is imbued with transcendental significance. The same is true of new reproductive technologies. *In vitro* fertilization, artificial insemination by donor, or surrogate motherhood may elicit one kind of moral judgment from a "Yuppie", quite another from an Orthodox Jew, a Roman Catholic, or a feminist. Such moral judgments, Engelhardt contends, are not easily translatable (and perhaps not even meaningful) across cultural boundaries. Because reason alone cannot tell us which the morally correct vision is, he concludes that "(t)here appears to be an insurmountable barrier to consensus formation about the moral significance of the new reproductive technologies" ([10],

p. 10).

Engelhardt's view seems to be confirmed by the experiences of broadly based government committees charged with the task of making recommendations on the use of reproductive technologies. After a two year inquiry, Mary Warnock, the chairperson of the British Committee of Inquiry into Human Fertilisation and Embryology, had come to the conclusion that "common morality" is a myth. She found that especially in an area as radically new as the one she was asked to inquire into, "the notion that there is a consensus morality ... is ... untenable" ([26], p. xi). Subsequent government inquiries into *in vitro* fertilization and related technologies bear out her observations [12].

If it is correct that modern western societies are characterized by a plurality of moral visions, then ethical theories of consensus are faced with a problem when it comes to applying them in practice. To the extent that the consent of all those affected is necessary for legitimizing particular ethical norms, it would seem impossible to establish *any* authoritative moral norms on issues on which members of a community are deeply divided. For, as we have already seen, the simple fact that a *majority* of those affected might agree on a matter does not impart moral legitimacy or authority.

This raises again the issue of the gap between "hypothetical" and "real" consent, touched on at the beginning of this Section. To begin, let us look at the issue in terms of contractualist theories of ethics, of which John Rawls' *A Theory of Justice* [24] is probably the best known one. Since the agreements reached in the "original position", by people who are motivated in very specific and limited ways, are "hypothetical" rather than "real", it is not clear why differently motivated actual people should regard those agreements or procedures as binding. After all, people do not normally consider themselves to be bound by agreements they did not make, or by principles or norms to which they did not consent. The most plausible response to this objection is that the conditions under which consensus is reached are *fair* or *impartial*, that is, that arbitrary and morally unjustifiable differences between the parties to the agreement have been removed to arrive at certain universalizable principles or norms. If "real" people refuse to be bound by the norms selected under such fair or impartial conditions, it is only because they are seeking to further their own interests, even if this is unfair or unjust. In other words, the norms agreed to under "ideal" conditions *would* find universal acceptance if people acted justly and fairly, and were to adopt an impartial point of view.

This makes it crucial to ask whether the principles and norms proposed by a particular contractualist theory of ethics are impartial and acceptable from "the moral point of view". While contractualist theories of ethics come in various forms, they generally have one thing in common: the idea that isolated, self-interested individuals seek to maximize their interests by making binding contractual or promissory arrangements. These contractual arrangements have a cost, but this is presumed to be outweighed by a greater gain for all those who are party to the contract. Contractualist theories of ethics thus rest on the implicit assumption that ethics is concerned with maximizing the good of self-interested rational agents, and the related assumption that the principles agreed on, by these agents, are impartial and acceptable from "the moral point of view". This has been disputed. Soon after John Rawls' *A Theory of Justice* had been published, Stuart Hampshire charged that the theory, far from providing a universally acceptable view of justice and rights, admirably expressed the ideas of the British Labour Party [15], and R.M. Hare argued that Rawls' "principles of justice" improperly favor the participants to the contract – existing rational agents – as against potential and possible people, while being altogether silent on non-human animals [17]. This is not the place to examine in detail the various claims and counter-claims. Nor is it necessary to do so. The basic problem posed for contractualist theories of ethics can be stated quite simply: Even if contractualist theories of ethics were impartial or fair as far as the principles governing conduct between the contracting agents are concerned, there is no universal agreement that morality is adequately characterized as a system which seeks to maximize the broadly construed interests of those who are party to the contract. Or, to put the point somewhat differently, while a contractualist theory such as Rawls' may give us *one* "moral point if view", this point of view is not the only possible one. Morality is frequently conceived of as much broader than that.

It is, of course, true that only moral agents can sign contracts, make agreements, and give and keep promises. However, even though one may want to accept that consensual arrangements between persons can give rise to moral obligations, this does not mean that ethics is exhausted by these contractual obligations. Morality may also quite properly be conceived of as a system which, for example, requires us to give equal consideration to the interests of all those affected by what we do

– where those affected need not be already existing rational agents, but could be *in vitro* embryos, the people who may or may not be born depending on whether, say, surrogate motherhood is allowed as a permissible option, or they could be sentient non-human animals. Because contractualist theories give no direct "moral standing" to either potential or possible persons, or to non-human animals, they are faced with some difficult problems. In the field of new reproductive technologies, this would entail that *in vitro* embryos, fetuses and those who might be born depending on whether particular technologies such as embryo donation, gene therapy or surrogacy will be adopted, have no "moral standing" – not because substantive reasons have been provided for their exclusion from the moral sphere, but rather because the heuristic device (a hypothetical contractual agreement between existing self-interested rational agents) has excluded them right from the beginning.

Taken to its logical conclusion, this would seem to entail that scientists would, for example, be free to experiment, with the parents' consent, on human *in vitro* embryos and those who have not attained "the age of reason", the condition which Rawls regards as a precondition for being party to the negotiations in the "original position" ([24], p. 146). These conclusions conflict with "ordinary morality" and with competing theories of ethics. It will come as little surprise, therefore, that contractualists are attempting to mute the stark conclusions of their theories. However, even if it were possible to overcome these difficulties as far as human beings are concerned (those who are party to the contract could, perhaps, imagine themselves as an *in vitro* embryo, or as the being into which the embryo would develop, and prudently decide that they would not want to be treated in certain ways), it is difficult to see how non-human animals, including animal/human hybrids, the not too fanciful products of new reproductive technologies, could find a niche in the contractualist's moral scheme. Here, it would seem, the only strategy open to a contractualist would be to view any obligations to such creatures as indirect obligations to humans. I shall return to this point in a moment, in my discussion of H. Tristram Engelhardt's consensual moral framework.

Enough has been said, it seems to me, to show that the claim that contractualist theories of ethics are providing us with a set of universally acceptable moral principles is misplaced. In taking hypothetical contracts between self-interested, rational agents as their starting point, they are foisting on us a substantive view of morality, unlikely to be acceptable to anyone who does not already share the contractualist's particular vision of "the good life". If contractualist theories thus fail to provide a universally acceptable framework for the resolution of moral disputes, the time has come to look at alternative approaches.

IV.

Some theories of consensus, taking respect for freeedom as their starting point, attempt to solve the problem of moral conflict by explicitly dividing morality into two parts – a public part and a private part. These theories grant people a private moral sphere in which they are free to shape their own lives in accordance with their particular vision of the good life, and insist on consensus only in regard to the normative framework employed [5]. This device, then, might help us to resolve the conflict between the Archbishop of Liverpool and Baroness Summerskill. Even if they cannot agree on the morality of *in vitro* fertilization procedures, they might nonetheless agree on procedures or norms which will allow them to resolve the ethical conflict in a mutually satisfactory way.

As our discussion of contractualist theories has suggested, to be universally acceptable a proposed principle, norm, or procedure would not only need to be able to meet the criterion of impartiality or fairness when viewed from within a particular moral scheme, such as Rawls', it would also have to be *neutral* between different moral schemes – for example, between a Rawlsian conception of morality and a utilitarian or liberal one: it must be acceptable to a traditional Christian, to a Yuppie and a social reformer alike. The question is whether such a neutral principle can be found – a principle which does not already harbor within itself a particular vision of "the good life". While the principle of universalizability, formally stated, would be neutral in the required sense, those who have attempted to derive concrete ethical principles or norms from it have invariably been accused of smuggling their own particular vision of morality into the picture.

This apparent inability to derive a content-full ethical theory from reason alone has recently led to a shift in focus. Even if it does not appear possible to develop a generally acceptable ethical theory, or to find universally acceptable concrete ethical principles or norms, might it not be possible to discover a neutral *procedure* for the resolution of moral disputes? H. Tristram Engelhardt's answer is "yes". He believes that a neutral procedural basis for ethics can be found "in the very nature of ethics", which he understands minimally as "an alternative to force in resolving moral controversies" ([11], p. 41). Let us examine his argument. Engelhardt holds that his minimum notion of ethics "commits one to no particular concrete moral view of the good life" ([11], p. 41). This seems to be correct. The definition of ethics as an "alternative to force in resolving moral controversies" has no specific content (other than to rule out forceful procedures of conflict solution), and does not tell us, for example, what the moral controversy is about, or how it is to be resolved. It is thus compatible with a wide variety of different and conflicting ethical views. But is it possible to derive a content-free procedural ethics from this minimal definition of ethics? Engelhardt believes it is. He argues that this understanding of ethics contains "as a necessary condition (...) the requirement to respect the freedom of the participants in a moral controversy" ([11], p. 42). From this minimum notion of ethics and the necessary condition of "respect for freedom", Engelhardt then arrives at the conclusion that individuals should be free to do as they wish, as long as they respect the equal freedom of others ([11], p. 45). This means that people should be able to use reproductive technologies as they wish, "as long as those who disagree are not constrained to collaborate with them." Those who disagree, he continues,

> should be at liberty peaceably to announce the damnation, particular and general, of all who use these technologies. In the absence of the possibility of a concrete consensus with regard to the moral significance of the human reproductive technologies, there should be freedom to go to hell as one wants, and to damn those who appear headed in that direction ([10], p. 35).

Now, this is clearly a set of substantive conclusions. How did Engelhardt derive them from his minimum notion of ethics as an "alternative to force in resolving moral controversies"? The answer lies in the subsequent adoption of the principle of "respect for the freedom of moral agents involved to do as they wish....". For while Engelhardt is correct when, in one of his formulations, he says that "[i]f one is interested in resolving moral controversies without recourse to force as the fundamental basis of agreement, then one will have to accept peaceable negotiation among members of the controversy as the process for attaining the resolution of concrete moral controversies" ([11], p. 41), he is wrong when he subsequently equates "respect for peaceable negotiation" with "respect for the freedom of others to do as they wish....". The first principle, but not the second, may be seen as a necessary condition for the peaceful resolution of moral disputes. "Respect for the freedom of others to do as they wish...." is a substantive principle that cannot be derived from Engelhardt's minimum notion of ethics, nor from the necessay procedural presupposition of peaceful negotiation. Engelhardt's principle of "peaceable negotiation" is akin to the "transcendental-pragmatic foundation" of ethics advanced by Karl-Otto Apel and other proponents of a "discourse ethics", according to which one cannot consistently reject rational discourse as a method for reaching and grounding normative conclusions [2]. But, as is well-recognized, such transcendental-pragmatic foundations of ethics cannot give rise to concrete practical norms – other than the implicit norm that consensus should be sought discursively [6]. This means that the only necessary condition Engelhardt can derive from his minimum notion of ethics is that we continue to talk to each other in our joint quest for a peaceful and mutually acceptable solution – not that we respect the freedom of others to do as they wish, within the constraint of respecting the like freedom of others. The principle that we should respect the freedom of others to do as they wish is no longer a procedural principle but a content-full vision of "the good life" – a life in which "autonomy" trumps other morally relevant considerations. The next section will amplify this point.

V.

We already noted that contractualist theories of ethics face problems with regard to the treatment of possible and potential persons, and of non-rational beings. I want to sharpen these points by focusing on some of the practical implications of Engelhardt's proposed normative framework.

If people ought to be free, as Engelhardt suggests, to do as they wish, as long as they do respect the like freedom of others, this would seem to put those who are not (yet) autonomous moral agents outside the moral sphere. To return to the contractualists' problem regarding the treatment of non-human animals, Engelhardt's liberal framework would seem to entail that you would be free gratuitously to torture non-human animals and those humans who are not persons, but that I – who would be free, as Engelhardt suggests, to publicly condemn what you do – must not use force to prevent you from doing what you do. Of course, you and I and members of society in general might agree that we ought to respect each others' feelings in these matters, or ought to recognize some duty of beneficence to non-rational beings. To the extent, however, that substantive principles such as beneficence are subordinate, in the moral framework proposed, to autonomy (recall that the framework is intended to secure for autonomous individuals a private sphere, that is, to protect them against the imposition of a concrete vision of "the good life" from outside), this means that non-rational beings have no "moral standing" in their own right, and that there is nothing intrinsically right or wrong in treating them in certain ways – provided that such treatment does not infringe the autonomy of others.

This response avoids what is essentially a *moral* question by turning it into a subjective value judgment, or into a procedural issue of consent. This is wrong-headed. To make a moral judgment is to make a universal claim – it is to prescribe for all relevantly similar situations, and not simply to state one's particular likes and dislikes. Nor can moral judgments always be reduced to questions of consent. When I am objecting to the gratuitous torturing of your dog, then I am not (or not primarily) objecting to it because the thought of torture evokes feelings of horror in me, and the fact that your consent has not been sought is of only peripheral concern. Rather, I am objecting to the torturing of your dog because I think that it is wrong to torture dogs and other sentient creatures because of what it does to *them*, not because of what it does to you or to me. In other words, I believe that non-rational, sentient creatures have "moral standing" and should not be excluded from the moral sphere.

This has obvious relevance for the possibility of achieving consensus on the use of reproductive technologies and the related topic of embryo experimentation. When Roman Catholics or members of the Right to Life movement are saying "embryo experimentation is wrong", they are making a *moral* claim. They are not particularly concerned to find out whether the parents' consent has been obtained, nor do they primarily object to the practice because it fills them with horror. Rather, they want destructive embryo experimentation stopped because they believe that the intentional termination of all innocent human life is wrong, or because they believe that every human being, from the moment of conception onwards, has a right to life. The same is true of many other objections to new reproductive technologies as well. People object to these technologies because they regard them as *morally* wrong. If these moral convictions are strongly held and if something of great moral significance, from the proponents' point of view, is at stake – for example, an embryo's "right to life" – then they are unlikely to give their assent to a moral framework which would require them to stand by while the most serious moral wrongs are committed by others.

Engelhardt's procedural approach to ethics will thus not be able to solve the problem of practical ethical conflict. The reason is not only that he has offered us a substantive account of ethics, rather than a neutral procedure; the problem runs much deeper than that. *Any* procedural account of ethics – and that includes the previously mentioned discourse ethics of Karl-Otto Apel and Jürgen Habermas ([2]; [13]; [14]) will either be empty – that is, offer us a contentless neutral procedure for the resolution of ethical conflict only – or it will have to recognize that the *content* of these discussions is itself in need of justification. In other words, before the discursive procedure can be applied in practice, one will need to know what the discourse is to be about, that is, who or what is to be included in the moral sphere – for example, every human being, all rational agents, all sentient beings, or all living things? The answer to those questions cannot be found within the procedural discursive approach to ethics itself. For while it is obvious that only rational agents can participate in ethical discourse, it does not follow from this that theirs are the only interests that count. The dilemma for consensus theories of ethics is that once a decision has been made to draw the moral boundary around one or the other of these and other substantive categories, the "neutrality" of the procedure has been lost. A substantive vision of "the good life" has been introduced which is not likely to be universally acceptable.

VI.

Next, I want to offer some positive reasons why a normative framework for the use and implementation of reproductive technologies should not be limited to questions of autonomy and consent. Kurt Bayertz has already shown why such frameworks, applied to ethical issues raised by reproductive technologies, are faced with some serious problems insofar as it is not possible to obtain the prior consent of the human beings to be conceived ([5], pp. 53–54). I want to raise two related problems to show that there are good reasons why we should reject the liberal conclusion that there should be no restrictions on the use of reproductive technologies and that the state must, in the words of Engelhardt, "abandon attempts to regulate reproductive technologies, other than to ensure that citizens are protected against fraud and other varieties of unconsented to harm and coercion" ([10], p. 20).

The first problem is a variation of what we might call Derek Parfit's "Baby Problems" ([22]; [23], pp. 358–9). It is intended to illustrate the possible implications of an unfettered principle of reproductive freedom. Imagine a couple on an IVF program. A test has become available which shows that one of the woman's eggs, as yet unfertilized in the petri dish, has a rare defect which would result in any child being born from it being severely handicapped. The child would die, after much suffering, before it is one year old. There would be no problem in collecting another egg during the woman's next cycle, and it is quite unlikely that such an egg would have the same defect. Assume that the parents, in full knowledge of the facts, decide to have the present egg fertilized and implanted. Should parents be free to make use of technologically assisted reproduction in this way?

"Respect for reproductive freedom" would entail that the parents should be free to bring a severely handicapped child into the world, even if this results in considerable suffering to the child. Because the child can obviously not give, or withhold, its consent to be brought into existence, those consensus theorists who are worried by the conclusion that people ought to be at liberty to inflict harm on those unable to consent, might want to add another ethical principle to their framework – the principle of non-maleficence. Such a principle could be conceived of as stating that it is, other things being equal, wrong to knowingly bring a child into the world for whom it would have been better if it had never been born.

This response would, however, pose a serious problem for the moral framework itself. Not only would the principle of non-maleficence now sometimes trump autonomy (thereby imposing a content-full vision of "the good life" on morally autonomous agents), but the acceptance of a principle of non-maleficence might also require the acceptance of a parallel principle of beneficence. In other words, if the *harm* a future child will experience is a reason for *not* bringing it into existence, might it not be argued that the *benefits* a future child would experience must, by parity of reasoning, count as a reason *for* bringing it into the world? If this were admitted, autonomous moral agents would not only be obliged to refrain from doing harm, they would now also be required to live by the much more arduous requirement of promoting the good.

But let us assume, at least for the moment, that it is possible to hold that we have a duty to refrain from not bringing utterly miserable beings into the world, but that we do not have a parallel duty to create happy beings. Now consider another couple on an IVF program: A test has become available which shows that the man's frozen sperm is defective. Any child conceived from it would be severely handicapped, but would still enjoy life and not think that she has been harmed by being brought into existence. Tests have shown that the man's medical condition, which resulted in the sperm being defective, has passed and any sperm collected now would be normal. The parents decide to use the frozen sperm, and nine months later a severely handicapped child is born.

The parents' behavior would not be ruled out by the principle of non-maleficence. For if the parents had used fresh, rather than frozen sperm, the child born from the alternative set of gametes would have been a *different* child, and the present child would not have existed at all. Since life, for the handicapped child, is still of positive value, the child has not been harmed by having been brought into the world. But is it obvious that the parents should, in the context of technologically assisted reproduction, be at liberty to choose an option which is clearly not "for the best"?

Now, it might be objected that parents, wanting the best for their children, are quite unlikely to engage in the kind of reckless reproductive behavior just outlined. This may well be true most of the time and, in the context of "natural" reproduction, there would have been little reason why the state should have stepped in to interfere with peoples' reproductive freedom. New reproductive technologies have, however, opened new windows of knowledge and opportunity. Technologically assisted reproduction is, in distinction from "natural" reproduction, not a private process, and human gametes and embryos have become accessible to observation and intervention – presenting not only parents but also doctors, scientists and policy makers with new options, responsibilities and temptations. In these changed circumstances it may no longer be sufficient to rely on the principles of "respect for reproductive freedom" and non-maleficence to guide reproductive choices.

The following scenario will illustrate the point. In another of his examples, Derek Parfit asks us to think of a nation choosing between two different industrial policies ([23], pp. 361–364). One policy will have disastrous ecological consequences, entailing early death from cancer for a million people, many generations from now. Depending on which policy is adopted, different people will come to exist, due to the different life-styles, choices of partners, and so on. Unless we assume that the lives of the people now dying of cancer have, on the whole, been so bad that it would have been better if they had never been born, we cannot say that they have been harmed by the actions of their forebears. For had the non-polluting policy been adopted, these people would not have been born. This means that the addition of the principle of non-maleficence to the liberal framework cannot yield the judgment that the non-polluting policy ought to have been adopted. It is not difficult to imagine parallels in technologically assisted reproduction. Because new reproductive technologies, especially when coupled with recent developments in the field of genetics, have afforded us many new options, it is quite conceivable that a future society may be faced with a choice between two different genetic policies, in much the same way as Derek Parfit's imaginary society was faced with a choice between two different industrial policies, with similar results.

These examples illustrate, it seems to me, that public policy decisions in the area of technologically assisted reproduction (as indeed in many other areas) must not be based on the principle of autonomy alone. Parfit's problems raise moral issues which cannot be handled by consensus theories because these issues are logically prior to the setting up of the consensus procedure. They are *substantive* issues that cannot be avoided by any purely procedural suggestions.

VII.

I have approached the problem of ethical conflict and consensus from various perspectives. The inevitable conclusion is that it is unlikely that consensus on the implementation and use of technologically assisted reproduction can be achieved. This leaves us in an apparent quandary – for in a situation such as this it is not an option for policy makers simply to do nothing. The decision not to allow the implementation and use of these technologies would be as unacceptable to certain sections of the community – the infertile and research scientists engaged in embryo experimentation, for example – as would be the decision to prohibit their use. What, then, are societies faced with ethical conflict to do?

The first thing to note is that there can be, as Jonathan Moreno has pointed out, various kinds or levels of moral conflict and consensus ([21], p. 395). These different levels were already implicit in my earlier discussion. There can be conflict or agreement on the underlying principles or theory, there can be conflict or agreement on a proposed solution, and there can be conflict and agreement on other levels in between. The question is what kind of consensus should we aim for on a public policy level?

There may be much in Bruce Ackerman's suggestion that, in our quest for consensus, we should exercise what he calls "conversational restraint" ([1], p. 15). In other words, if you and I are taking fundamentally different views of "the good life", then it may be quite appropriate to leave the question of moral ideals off the agenda. I will not attempt to turn you into an atheist, provided you will not try to convince me that I should become a Buddhist or a Roman Catholic. Having constrained our discourse in this way, we may, Ackerman thinks, be able to use dialogue for pragmatically productive purposes: to identify normative premises that those participating in the dialogue can reasonably accept ([1], p. 15). The aim in such consensus-oriented discourse is not to discover "the ultimate truth", but rather to provide members of society with a "way of reasonably responding to their continuing moral disagreement" ([1], p. 19). This presupposes that reason and argument have *some* role to play in ethics. If one were to abandon this premise and believe that ethics is entirely a matter of subjective feelings or intuitions (where one person's intuitions are as good as those of any other), then it is unlikely that much would be gained by our engaging in dialogue or discourse on ethical matters.

But there is – as Socrates is not the only one to remind us – an important connection between discourse and the good or examined life for human beings. While discourse, talking to those who profess to know, may not furnish us with the "truth", it may nonetheless tell us where they, or we, have gone wrong. And, of course, discourse may also discover areas of agreement. While Mary Warnock had, as previously noted, remarked that "'common morality is a myth", she nonetheless noted that she was "more impressed by the extent of moral agreement than of disagreement among members of the committee, especially considering the many different professions, religions and races" of those who served on her committee ([26], p. x). In other words, discourse will often reveal that we tend to overestimate moral disagreement. Ethics is not an arbitrary series of different things to different people, and even the most difficult ethical issues are still amenable to reason and discussion.

If this is correct, the next question must be this: how is such discourse to be implemented in pluralist democratic societies? Ideally, all members of society should be able to engage in it, but this is clearly not possible in large-scale modern societies. As we have repeatedly noted, opinion polls cannot solve the problem. For even if a majority of the population were to agree on, say, the morality of surrogacy or embryo experimentation, this would not furnish the government with universal assent. There are, however, also more pragmatic reasons why one would not want to leave public policy decisions on complex matters, such as the implementation and use of new reproductive technologies, for direct decision by the general population. Minimally, one would want to be reassured that those whose opinion is sought have a sound understanding of all the facts of the situation. But this assumption cannot always be made when we are dealing with matters as complex as those before us. Take the issue of embryo experimentation. If *in vitro* embryos could feel pain, this would be an important reason against performing painful experiments on them. But early embryos, consisting of no more than a few cells, do not have a central nervous system and have no sensory awareness. It is therefore disturbing to find that a university study showed that of 130 undergraduate students some 44% thought that early embryos had a brain; and some 21% thought that these embryos "could feel" [27]. Such factually wrong beliefs, probably even more prevalent in the general community, may well inform people's moral judgments on issues such as embryo experimentation. To the extent that these judgments rely on false beliefs, they must be discounted by those charged with making public policy decisions.

This means that if the community as a whole were to decide directly on special issues that fall outside the normal competence and knowledge base of members of the community, then the community would need to be educated first before it could properly be asked to make a judgment on those issues. While such educational goals are laudable and should be encouraged when important issues are at stake, they are often difficult to achieve. Issues raised by the implementation and use of new reproductive technologies presuppose not only knowledge of the relevant technical facts, but also interdisciplinary research and analysis of the complex ethical, social and public policy questions raised by them.

It is for reasons such as these that modern democracies function as representative democracies: members of the community hand over to the elected representative the task of deciding on issues on which the community as a whole is not, and perhaps cannot be, adequately informed. Modern democracies thus rest on the principle of representation: because legislators, in distinction from ordinary members of the community, have the time and resources to acquire the relevant expertise in the area under consideration, they act as representatives for the community, rather than simply reflecting the views of those who elect them. As Edmund Burke put it in his speech to his electors in Bristol: "Your representative owes you, not his industry only, but his judgment; and he betrays instead of serving you if he sacrifices it to your opinion" [7]. When an issue is particularly complex or sensitive, a government will often decide to hand it to a broadly based specialist committee. This takes the principle of representative government one step further. The committee can provide an additional level of specialization and division of labor and, via the elected representatives, represent the people. More particularly, a permanent committee – a national bioethics committee, for example – could function as the locus of ethical discourse par excellence.

What I am suggesting, then, is this: that governments, intent on establishing a moral framework justified in terms of consent on the implementation and use of reproductive technologies, set up a national consensus-oriented bioethics committee to advise it on the ethical and public policy issues involved. The assumption would be that the government of the day will accept the committee's recommendations. While a national bioethics committee would not be the ultimate authority on moral truth, it would be in *authority* in the sense of having been set up by universal assent to act on society's behalf.

Such committees might, of course, be subject to various distorting influences ([20], pp. 416–17). Moreover, their conclusions will not please everyone – either because people disagree with the conclusions on substantive grounds, or because they think that the methodology employed is flawed [18]. For this reason it is important that committees do not simply state their views, but present reasoned arguments for their conclusions. In this way, it will be possible for those who disagree with them – and this will include members of the public – to enter the discourse, thereby contributing to the debate.

Ethics committees should not necessarily be seen as a microcosm of pluralist societies, but rather as a locus for the exercise of expertise and intelligence. This means that governments should ensure that all the relevant disciplines involved should be represented on the committee, to lend their expertise to the clarification of the issue in question. This should include philosophical or ethical expertise. Such expertise should not be conceived of as the kind of expertise that will necessarily help the committee to find "the truth", but rather as the kind of expertise that might prevent it from proceeding down one of the many paths that lead to error. In other words, the expertise I have in mind is not that of a Philosopher King, but rather that of being able to reason well and to detect errors in one's own and others' ethical thinking, familiarity with the different ethical theories, with moral concepts and the various approaches to ethics ([25], pp. 199–201).

Of course, also philosophers have particular visions of "the good life", and Mary Warnock's particular vision that morally dependent harms – that is, unreflective "outrage and shock" [16] – should be given moral weight, has undoubtedly colored the conclusions of her committee ([19]; [18]). It is difficult to see how this can be avoided. The best we can, perhaps, do in such situations is to ask philosophers who chair committees to state and defend their fundamental philosophical and ethical outlooks, so that these fundamental presuppositions, together with the more practical recommendations, can be tested in public discourse.

My own view is – and it is one which would require a much more extensive defense than I can provide here – that feelings of "outrage and shock" should not influence public policy making. If they were to be given proper weight, we might as well do away with committees, and instead of having reasoned discourse, informed argument and debate, we would be better advised to simply sample peoples' prereflective views by opinion polls. Such views would, however, have no authoritative force – for the reasons already discussed. The reasoned conclusions of a properly constituted and functioning committee, on the other hand, do have this force. They would have authoritative force not only because members of society have given their consent to resolve ethical conflict in this way, but also because the process of rational dialogue and discourse itself imbues the committee's conclusions with some authoritative force. This authoritative force is, however, tentative and fallible; it can be revoked when good reasons are found which better support an alternative conclusion.

National ethics committees will not be perfect, and I am probably not quite as confident about their efficacy as my last few pages suggest. Nonetheless, such committees are perhaps the best we can do in our quest to achieve consensus.

I don't know whether the Archbishop of Liverpool and Baroness Summerskill are still disagreeing on the morality of *in vitro* fertilization procedures. But if they do, it will perhaps not be too much to hope that they, like others unable to show that their particular vision of "the good life" is correct, will agree to the establishment of a national consensus-oriented bioethics committee. While this committee may not present them with "the truth", as they see it, they will, hopefully, be able to live with its reasoned conclusions.

Centre for Human Bioethics
Monash University
Clayton, Victoria, Australia

NOTES

* This article has greatly benefited from the papers presented at the 1990 Bielefeld conference "Technische Eingriffe in die menschliche Reproduktion: Perspektiven eines moralischen Konsenses", and the subsequent discussions. If I have not always been able to attribute an idea or a point to a particular person, I hope I will be forgiven. Instead, I would like to acknowledge a universal debt.

BIBLIOGRAPHY

1. Ackerman, B.: 1989, 'Why Dialogue', *The Journal of Philosophy* LXXXVI, 5–22.
2. Apel, K.-O.: 1988, *Diskurs und Verantwortung. Das Problem des Übergangs zur postkonventionellen Moral,* Suhrkamp, Frankfurt/M.
3. Aristotle: 1963, *Ethica Nicomachea,* in *The Works of Aristotle,* vol. IX, translated into English under the Editorship of W.D. Ross, Oxford University Press, London.
4. Aristotle: 1941, *The Politics*, in *The Basic Works of Aristotle,* ed. and with an Introduction by Richard McKeon, Random House, New York.
5. Bayertz, K.: 1994, 'The Concept of Moral Consensus. Philosophical Reflections', in this volume, pp. 41–57.
6. Boehler D., and Matheis A.: 1991, 'Töten als Therapie? – "Praktische Ethik" des Nutzenkalküls versus Diskursethik als kommunikative Verantwortungsethik', *Ethik und Sozialwissenschaften – Streitforum für Erwägungskultur* 2, 361–375.
7. Burke, E.: 1883–90, 'Speech to the Electors of Bristol, November 3, 1774', in *Works*, vol. I, Bohn, London, pp. 446–7, as cited in P. Singer and D. Wells: 1984, *The Reproduction Revolution – New Ways of Making Babies*, Oxford University Press, Oxford, New York, Melbourne, p. 195.

8. Devlin, P.: 1968, *The Enforcement of Morals*, Oxford University Press, London.
9. Edwards, R. and Steptoe, P.: 1981, *A Matter of Life*, Sphere, London.
10. Engelhardt, H.T., Jr.: 1994, 'Consensus: How Much Can We Hope for? A Conceptual Exploration Illustrated by Recent Debates Regarding the Use of Human Reproductive Technologies', in this volume, pp. 19–40.
11. Engelhardt, H.T., Jr.: 1986, *The Foundations of Bioethics,* Oxford University Press, New York, Oxford.
12. Gaze, B. and Kasimba, P.: 1990, 'Embryo Experimentation: The path and problems of legislation in Victoria', in P. Singer, H. Kuhse *et al.* (eds.), *Embryo Experimentation*, Cambridge University Press, Cambridge, pp. 202–212.
13. Habermas, J.: 1975, *Legitimation Crisis*, Beacon Press, Boston/Mass.
14. Habermas, J.: 1983, 'Diskursethik – Notizen zu einem Begründungsprogramm', in J. Habermas, *Moralbewußtsein und kommunikatives Handeln*, Suhrkamp, Frankfurt/M., pp. 53–125.
15. Hampshire, S.: 1972, 'A New Philosophy of the Just Society', in *New York Review of Books*, February 28, as cited by R.M. Hare: 1989, 'Arguing about Rights' in R.M. Hare: *Essays on Political Morality*, Clarendon Press, Oxford, p. 106.
16. Hampshire, S.: 1972, 'Morality and Pessimism', reprinted in S. Hampshire: 1978, *Public and Private Morality*, Cambridge University Press, Cambridge, as cited by R.M. Hare: 1990, 'Public Policy in a Pluralist Society', in P. Singer, H. Kuhse *et al.* (eds.), *Embryo Experimentation*, Cambridge University Press, Cambridge, p. 189.
17. Hare, R.M.: 1989, 'Rawls' Theory of Justice', in R.M. Hare, *Essays in Ethical Theory*, Clarendon Press, Oxford, pp. 143–174.
18. Hare, R.M.: 1990, 'Public Policy in a Pluralist Society', in P. Singer, H. Kuhse *et al.* (eds.), *Embryo Experimentation*, Cambridge University Press, Cambridge, pp. 183–194.
19. Lockwood, M.: 1985, 'The Warnock Report: a philosophical appraisal', in M. Lockwood (ed.), *Moral Dilemmas in Modern Medicine*, Oxford University Press, Oxford, New York, pp. 155–186.
20. Moreno, J.: 1988, 'Ethics by Committee: The Moral Authority of Consensus', *The Journal of Medicine and Philosophy* 13, 411–432.
21. Moreno, J.: 1991, 'Consensus, Contracts, and Committees', *The Journal of Medicine and Philosophy* 16, 393–408.
22. Parfit, D.: 1976, 'Rights, Interests, and Possible People', in S. Gorovitz *et al.* (eds.), *Moral Problems in Medicine*, Prentice Hall, Englewood Cliffs, N.J., pp. 369 – 375.
23. Parfit, D.: 1984, *Reasons and Persons,* Clarendon Press, Oxford.
24. Rawls, J.: 1971, *A Theory of Justice*, Oxford University Press, Oxford.
25. Singer, P. and Wells, D.: 1984, *The Reproduction Revolution – New Ways of Making Babies*, Oxford University Press, Oxford, New York, Melbourne.
26. Warnock, M.: 1985, *A Question of Life – The Warnock Report on Human Fertilisation and Embryology*, Blackwell, Oxford.
27. Wood, C. *et al.*: n. d., 'A survey of Monash University undergraduates' knowledge and attitudes concerning the early human embryo', Department of Obstetrics and Gynaecology, Queen Victoria Medical Centre, Melbourne, unpublished paper.

PART TWO

CONSENSUS IN LAW AND POLITICS

WOLF-MICHAEL CATENHUSEN

PROBLEMS INVOLVED IN ACHIEVING A POLICY CONSENSUS ON ISSUES RELATED TO REPRODUCTIVE MEDICINE

Since the 1970s, the field of reproductive medicine has seen the application of new technologies, many of which were originally developed for animal breeding, to humans. The reproductive technologies of in vitro fertilization (IVF) and embryo transfer (ET) are often associated with many other biological methods, such as cloning or embryo splitting, the deep freezing of sperm cells, ova or fertilized ova, the use of ova and sperm of donors, the assistance of surrogate mothers, and implantation diagnostics. There are also the techniques that genetic engineering has spawned for identifying, studying and manipulating the hereditary traits of human beings at the DNA level. This broad array of methods poses fundamental challenges to the way in which we, both individually and as members of society as a whole, understand and grasp human beings and human reproduction. One issue is our future understanding of family and our understanding of parents' responsibilities to their children. The new technological means for influencing reproduction have provoked social controversies about these questions, in response to which legislators are called upon to act. At the same time, this whole matter is also inseparably linked with the issue of whether or not it is admissible in an ideologically pluralistic society to force ethical standards and value judgments upon all members of such a society, either with or without the help of the law.

Some democratic states have already begun enacting legislation on aspects of human reproduction: the different laws on abortion that have been passed in various countries regulate the possible decisions that can be taken when there is a conflict between the protection that unborn human life deserves to be afforded and the interests of a pregnant

Kurt Bayertz (ed.), The Concept of Moral Consensus, 99–107.

woman. For many years, our [German] family law has also defined the legal status of children in relation to their biological and social parents. Policymakers set standards here, but also respond to changes in societal values. This is evidenced, for instance, by the evolution in the status of illegitimate and extramarital children vis-à-vis their parents.

Democratic societies like that of the Federal Republic of Germany are characterized by ideological pluralism. A democratic state is a religiously and ideologically neutral state. It cannot require by law that its citizens adopt a self-contained system of values that adheres to a certain philosophy or ideology. On the other hand, in Germany a number of basic ethical standards, i.e., fundamental values that people were able to agree on, were incorporated in 1949 into the basic rights and state objectives of the new German Constitution, the Basic Law. For example, Article 1 of our Basic Law stipulates that human dignity is sacrosanct, and obliges the state and all of its bodies to protect and respect the sacrosanct nature of human dignity. At the same time, this article of the Basic Law declares that the human rights are inviolable and inalienable "as the basis of all human society, of peace, and of justice" [9]. Of course, the way in which human dignity is interpreted is also subject to processes of societal change. The Federal Constitutional Court has been called upon repeatedly to pass judgment on the constitutional implications of the impacts of new technologies on society and the corresponding responsibilities of policymakers. In its 1978 decision on the construction of a fast breeder reactor in Kalkar it expressed the view that it is the task of the state – i.e., of the legislature – to protect the basic rights set forth in the constitution against being endangered by new technologies ([4], 49, 89). This task, of course, also applies to the questions posed by the advances in reproductive medicine. These questions cannot be responded to solely by decisions on the part of the scientific community. Nor can it be in our interests to simply cope passively with the implications by adapting our basic values and our legal system to the new possibilities that have been created by reproductive medicine.

The first task facing policymakers in this regard is to provide adequate information about the state of scientific and technical developments. Only in this way will it be possible to involve large parts of society in the discussion on the opportunities and risks of reproductive medicine, motivating them to make inquiries and express their criticism, fears and hopes. Here, policymakers are thus also faced with the special task of doing all they can to make sure that science meets its obligations to society, namely to inform and enter into a dialogue about the societal implications of its discoveries; after all, modern biology has also been characterized by "the self-sustained dynamics of scientific progress and its practical implementation racing ahead of reflection on its reach and its possible consequences" ([11], p. 346). Technology assessment is there to portray, in as much detail as possible, the requirements and effects of new technologies, as well as feasible options for action. The Office of Technology Assessment of the United States Congress has also concerned itself for many years with the fields of biotechnology and biomedicine. It has commissioned a large number of studies to enable others to analyze the societal consequences of using biomedical techniques and to draw upon a large body of information for identifying the needs for social and political action (cf. [10]). In Germany the Benda Commission, which was constituted by the Federal Government and existed from 1983 to 1985, and the Enquete Commission on "Chances and Risks of Genetic Engineering" of the German Bundestag (1984–1986) have both contributed to this process [2].

It is also the task of policymakers to help work to achieve a consensus within society, if such is possible, as regards the issue of responsible and accountable application of the possibilities offered by modern reproductive medicine. "A pluralistic society that respects the range of possible opinions and must not ascribe to any single ideology or creed" requires a broad consensus for this, since "only those ethical convictions can be of relevance from the standpoint of constitutional law that are borne by a consensus spanning all of society" ([1], p. 216). Benda correctly points out that a constitutional system founded on respect for human dignity necessarily depends on the widest agreement possible on certain fundamental issues. This consensus is needed particularly urgently in connection with the difficult question as to whether or not and if so to what extent the state ought to wield the legal instruments at its disposal in order to codify ethical assessments of the possibilities of reproductive medicine.

To begin with, political endeavors to achieve a consensus must take the public's inquiries and fears having to do with the application and possible abuse of reproductive medicine seriously, and examine these fears and criticism in a dialogue between scientists and the public. In order to identify the chances of attaining a social consensus on how to deal with reproductive medicine, it must be possible for scientists to talk with the public and policymakers. The style and kind of dialogue about the opportunities and risks of reproductive medicine has also been a good indicator of the state of relations between scientists and the public in Germany. Without a doubt, the misuse of medical research on human beings during the Third Reich, which is a matter of historical record, has contributed to the Bundestag's willingness to pass legislation on the use of reproductive technology that runs contrary to our ideas on protection of human dignity. Nor is there any question that the lack of a tradition of scientific publications aimed at a broad public audience, as well as the underdeveloped willingness of scientific organizations to engage in a dialogue with critical members of the public, have strengthened the position of those in Germany who advocate legislative control of reproductive technology.

Where the question of a social consensus on issues having to do with reproductive medicine is concerned, policymakers must allow as many groups of society as possible to participate in a qualified manner in the dialogue, and grant them the opportunity to form and express their own opinions. It is not only important to organize bodies to directly discuss policy and lay the groundwork for decisions on it; it is also essential for women's groups, trade unions, political parties, environmentalist associations, churches, etc., to formulate their inquiries and take active part, via representatives, in the dialogue with scientists and policymakers, thus facilitating the response of policymakers to the issue of a social consensus. Social dialogues conducted for the purpose of identifying a possible consensus of the scope portrayed here take time. The formation of commissions of experts by the government and the Bundestag in 1983 and 1984 to deal with these issues was also a reaction to the questions and concerns of the public. At the same time, the participation in expert-commissions representatives of social groups has fostered the formation and expression of opinions in these social groups.

Fourteen years after the birth of the first test-tube baby, Louise Brown, parts of the public still nurse misgivings about the new methods of technically assisted reproduction. At the same time, Germany has been one of the world's first countries – following an intensive public discussion lasting a number of years in which political parties, the Bundestag, and the Federal Government have taken an active part since 1983 – to introduce legislation, in the form of the Law on Protection of Embryos (*Embryonenschutzgesetz*) [7], to regulate the field of reproductive medicine. It is an initiative whose underlying convictions are shared by all of the political parties of the Federal Republic of Germany. In the process, a broad social consensus has been achieved in Germany, constituting the basis for legislative solutions. The rejection of the Law on Protection of Embryos by the opposition parties, the SPD and the Greens, was justified by the desire for even more stringent restrictions on the use of reproductive medicine.

In the political discussion, two fundamental questions were at the top of the list for clarification:

- Is reproductive medicine, as a provider of technical support for human reproduction, socially tolerable and acceptable?
- Who in society should be responsible for taking decisions on issues related to reproductive medicine?

1. Is reproductive medicine, as a provider of technical support for human reproduction, socially tolerable and acceptable?

Many reasons have been given in democratic societies for rejecting the possibilities of reproductive medicine. Gena Corea's book *The Mother Machine* lends expression to the rejection of reproductive medicine that has been principally championed by the feminist movement [6]. She is opposed to the use of animal breeding techniques on human women, which, in her view, debased them to mere objects of a reproductive technology employed by men and physicians. Similarly, reproductive medicine is regarded as an instrument for exerting social pressure on women, who must already "suffer" when they have remained childless in spite of desiring to have children, to make them believe that their wish for children must be fulfilled in order to restore their feeling of self-worth. At the same time, it is a fact that of the 12–15% of married couples without children, about half of these couples would like to have a child ([11], p. 346). Many of these couples have therefore also attempted to fulfill their wish for a child of their own with medical assistance such as hormone treatments. In Germany, such treatments have long been included in the catalogue of medical services that are covered by the state health insurance organizations. This social situation has also contributed to the fact that the basic doubts expressed by Pope John Paul II against in vitro fertilization and embryo transfer have received so little attention in the public discussion about the ethics of reproductive medicine [5]. The pope rejects in vitro fertilization with the same arguments that have already been used against birth control: he condemns the separation of coitus and conception that is necessarily associated with these techniques. However, this view is no longer at all appropriate to sexual behavior in the highly developed industrialized nations. Nor is the rejection of reproductive medicine in Germany, for which varying arguments have been advanced, a viable basis for social repudiation of modern reproductive medicine. Moreover, if the state were for instance to prohibit married couples to be able to fulfill their wish for a biologically conceived child of their own with technical assistance, then this would definitely constitute a violation of the basic liberties guaranteed to our citizens by the German Constitution. At issue is also the possibility that physicians may be obligated to help when treatment is desired to overcome childlessness. Nonetheless, questions and criticism remain concerning the consequences for society of applying techniques from the field of reproductive medicine, embracing technical, social and legal aspects, and all of the political parties and many associations, including the German National Medical Association (*Bundesärztekammer*) agree that these require regulation. What is therefore now emerging in Germany at a relatively early date is a societal consensus to approach modern reproductive medicine cautiously, especially in the problem areas of heterologous fertilization, use of ova and embryos from donors, and surrogate motherhood. This conservative attitude had already been practiced by the German scientific community during the development of the IVF and ET methods, which was of course dependent on a large number of trials in which fertilized human egg cells were sacrificed.

2. Who in society should be responsible for taking decisions on issues related to reproductive medicine?

Professor Hans-Peter Wolff, who for many years has chaired the Central Scientific Advisory Council on issues of reproductive medicine of the German National Medical Association, emphasized in a speech held before the National German Medical Congress in 1985 that in vitro fertilization with embryo transfer has "created facts and visions of what is socially possible and biologically feasible that are running up against the limits of conventional morals and traditional law. They range from the involvement of third parties in reproduction and parenthood, which is already practiced and commercially utilized today, through the elimination of genetically defective material to utopian techniques like cloning, which entails the production of any desired number of identical individuals from a single fertilized ovum" ([11], p. 347). That same year, the German National Medical Association issued its "guidelines for the performance of in vitro fertilization (IVF) and embryo transfer (ET) as methods for treating human sterility" [3]. These guidelines represented an attempt to get a grip on the technical, social, legal and ethical problems that, in the view of the German medical community, were associated with IVF and ET, by addressing the issue of professional ethics.

Any attempt to regulate the application of reproductive medicine must necessarily address the potential undesired and unforeseeable consequences and manipulations from the actions of physicians. Regulation naturally also calls upon their professional community to provide answers. For another, in this situation the medical community wanted to maintain and defend its right to deal with such issues autonomously, as opposed to a general social and political right to do so. Professor Wolff left no doubt about this in his speech: "In the view of the Commission, it is up to the professional organization itself to form interdisciplinary commissions spanning different professions for the purpose of deliberating on and bindingly defining the limits of what is ethically and medically admissible, the scope for physicians' decisions, and the rules for practical performance of procedures. ... The activities of the legislature should be limited to adjustments of family and inheritance law to take account of the new situation, and to basic decisions on surrogate motherhood and on the commercial use of fertilized ova and embryos" ([11], p. 353). This conflict between the claim asserted by the scientific and medical community that it alone is entitled to regulate issues of importance to the actions of physicians on the one hand and the comprehensive regulatory claim of the legislature on the other has also influenced endeavours in other countries to achieve a consensus and regulation of the field of reproductive medicine at least as much as differences of opinion and differing values within society as regards the content of possible regulatory approaches.

In Germany, like in other European countries, it has taken over ten years for political decisions to be made on reproductive medicine. An advantageous consequence of this is that it has been possible to incorporate experience gained with and improvements to the methods of reproductive medicine in legislative decisions. Moreover, the number of children who have been conceived worldwide with IVF and ET is still relatively small. In addition, endeavours to achieve a social consensus on use of the new scientific and technical possibilities must consider that differing phases can occur during the course of public reactions: hopeful phases are common in the beginning, amplified by the extensive promises made by scientists; then come critical reactions and fears that are prompted or intensified by the emergence of practical problems of introduction (e.g., low success rates); and this leads to doubts as to the ability of human beings to responsibly cope with their new skills. This in turn gives rise to the hope that a society could spare itself abuse-related issues by prohibiting new scientific techniques at the start. However, having taken this stance, critics are quickly forced to admit to themselves that a society which cannot be trusted to prevent irresponsible use of reproductive medicine cannot be trusted to implement blanket prohibitions, either.

At the end of a nearly ten-year search for standards by which to measure the responsible practice of reproductive medicine in Germany, a broad political and social consensus has emerged on important issues. It consists for one of the rejection of using reproductive medicine in a way that can only lead to commercial, anonymous production of human life. For this reason, the German Law on Protection of Embryos has made it illegal under threat of punishment [7]:

- to commercially arrange for a surrogate mother, or for a physician to be involved in implantation of an embryo in a surrogate mother; and
- to trade in or donate fertilized human ova.

The German Law on Protection of Embryos furthermore prohibits the use of fertilized human ova for research purposes. It is for this reason that the Law on Protection of Embryos stipulates that only as many ova as are needed may be fertilized for an attempt to induce a pregnancy. This is intended to prevent in advance the creation of human embryos that could be used for research purposes. In connection with the political discussion in Germany on whether or not to prohibit embryo research, the question arose as to whether the claim made by the medical community was justified, namely that, in view of the many abortions performed in our society, it constitutes a schizophrenic moral perception to call for protection of embryos at the expense of gaining highly valuable knowledge from studying embryos. The broad consensus that has been achieved in Germany to prohibit embryo research is based first and foremost on the view, shared by all sides, that protection must be given to emerging human life starting from the moment in which a human ovum is fertilized. The conflict between the protection of emerging human life on the one hand and the interests and conflicts of a mother-to-be on the other, however, cannot be compared with the situation of a physician interested in obtaining research results who would like to weigh the protection of emerging human life against the priorities of his research interests. It is not acceptable for physicians to make decisions of their own on this. Rather, emerging human life must be protected against interventions that do not pursue therapeutic objectives. Indeed, this attitude has permitted many of those opposed to applying criminal law to pregnancy conflicts to agree to a prohibition of embryo research.

Member of the German Bundestag
Chairman of the Committee for Research, Technology and Technology Assessment
Bonn, Germany

BIBLIOGRAPHY

1. Benda, E.: 1985, 'Erprobung der Menschenwürde am Beispiel der Humangenetik', in R. Flöhl (ed.), *Genforschung – Fluch oder Segen*. J. Schweitzer, München, pp. 205–231.
2. Benda Kommission: 1985, *Bericht der Arbeitsgruppe In vitro-Fertilisation, Genomanalyse und Gentherapie*, Bonn.
3. Bundesärtzekammer: 1985, 'Richtlinien zur Durchführung von In vitro-Fertilisation (IVF) und Embryonentransfer (ET) als Behandlungsmethode der menschlichen Sterilität', in *Deutsches Ärzteblatt* 82, (22), 1690–1698.
4. Bundesverfassungsgericht (Federal Constitutional Court), *Entscheidungssammlung*, FRG.
5. Congregation for the Doctrine of the Faith: 1987, *Instruction on Respect for Human Life in its Origin and on the Dignity of Procreation*, Vatican City.
6. Corea, G.: 1986, *The Mother Machine. Reproductive Technologies from Artificial Insemination to Artificial Wombs,* Harper & Row, New York.
7. Embryonenschutzgesetz. Gesetz vom 13. 12. 1990, in *BGBl.* I, p. 2746.
8. Enquete-Kommission des Deutschen Bundestages, W.M. Catenhusen, H. Neumeister (eds.): 1987, *Chancen und Risiken der Gentechnologie. Dokumentation des Berichts an den Deutschen Bundestag*. J. Schweitzer, München.
9. *Grundgesetz für die BRD vom 23. Mai 1949* (German Constitution).
10. U.S. Congress, Office of Technology Assessment: 1984, *Human Gene Therapy – A Background Paper*, US Government Printing Office, Washington, D.C.
11. Wolff, H.P.: 1985, 'Ein Kind mit fünf Eltern ? Die Befruchtung außerhalb des Mutterleibes erfordert Richtlinien', in R. Flöhl (ed.), *Genforschung – Fluch oder Segen*. J. Schweitzer, München, pp. 346–353.

CARL WELLMAN

MORAL CONSENSUS AND THE LAW

New medical technologies typically bring with them new moral problems. This is abundantly and awkwardly apparent in the case of the recently developed human reproductive technologies. Do infertile couples have a moral right to medical assistance in reproduction? Does the donor in artificial insemination have any financial responsibility for the financial support of his child? Is it morally permissible to fertilize *in vitro* more ova than will be implanted in the female patient? Who ought to have custody of fertilized ova held in storage for possible future use? Ought the law to enforce surrogate motherhood contracts? Is surrogate motherhood itself morally permissible? Does a pregnant woman have a moral duty to submit to unwelcome medical treatment necessary for the health of her unborn child? Bioethicists and medical practitioners are in doubt about how these questions should be answered, and any proposed solution to these pressing moral problems will be highly controversial. Finding disagreement where concerted action of patient, physician, health care institution and public authorities is urgent, one longs for some consensus on these and similar moral issues.

But why this longing? Why, above and beyond the desire to avoid unpleasant disagreements, would one want a moral consensus concerning technical interventions in human reproduction? Any comprehensive discussion of the desirability of such a consensus would exceed the limits of any readable essay and my limited philosophical capacities. In any event, I doubt that any general answer is possible. One would want moral agreement between patient and attending physician to preserve the moral integrity of each, to ensure that the patient could consent to the advised medical treatment in good conscience and that no physician would be called upon to provide what she considers to be immoral treatment. A hospital ethics committee might prefer to reach its conclusions by consensus rather than majority rule because voting tends to polarize

Kurt Bayertz (ed.), The Concept of Moral Consensus, 109–121.

individuals in a manner detrimental to the candid expression of personal conviction and the mutual good-will so important to the reasonable discussion of delicate and momentous issues. Physicians should value a moral consensus among themselves concerning medical practices in order to maintain the mutual respect needed for professional collegiality and to enable the profession to discipline its own members on the basis of shared standards. At the same time, physicians will desire a moral consensus in the larger society concerning their medical practices to preserve the public image of their profession, an image conducive to their professional goal of serving that public. All of this suggests that one's reasons for wanting a moral consensus concerning the use of the new reproductive technologies will depend upon who one is and the context within which one is functioning. Accordingly, I shall limit my attention to the legal context, the only context about which I am even remotely competent to speak.

Why should the law *as such* want a moral consensus concerning technical interventions in human reproduction? The obvious reasons are to make legal regulation of biomedical research and treatment of human reproduction unnecessary. One would prefer not to regulate these areas of human conduct because here, as elsewhere, legal regulation has its costs. For a start, institutional resources are expended in the very process, often complex and extended, of making new laws, whether through legislation or judicial precedent and in enforcing laws once made. There is also a price to be paid in terms of the freedom of patients, medical care providers, human subjects and biomedical experimenters. Moreover, since laws must be formulated in advance and in general terms, they are somewhat inflexible over time and make little allowance for the special circumstances of particular cases. Hence, generally beneficial laws occasionally prove harmful, especially when they regulate new technologies where the risks and benefits are not yet reliably known. Therefore, one would prefer either to have no regulation or to have some sort of informal regulation, say by the practicing physicians themselves or institutional review boards, that could be better informed, more flexible, and less restrictive. But a society can safely leave important and potentially dangerous interventions without legal regulation only if there is a sufficient degree of moral consensus so that individuals can be expected to act morally without regulation or more informal regulation can be trusted to be morally enlightened.

Unfortunately, legal regulation may be necessary in areas of human conduct where liberty is often abused and important moral values are in jeopardy. Presumably the purpose of making and enforcing laws that require the performance of moral duties and prohibit morally wrong acts is to increase dutiful action and decrease wrongdoing. But enforcement alone is insufficient to achieve this purpose; most, not all, of those subject to the law must believe that the law is morally justified so that they will at least obey and at best support the law. The classic example of the inadequacy of mere enforcement in the United States was the prohibition of the manufacture and sale of alcoholic beverages; a more recent and relevant example was the frequency of illegal abortions, with all their attendant harms, before *Roe v. Wade*. Laws reflecting moral standards will be effective only if there is something approaching a moral consensus on those standards in the society.

There are also special contexts within the law where special reasons to want a moral consensus concerning technical interventions in human reproduction apply. Legislators will want something approaching a moral consensus among themselves in order to make practicable those institutional compromises necessary to enact statutes under majority rule and to exclude the danger that by voting for necessary legal regulations they will find themselves isolated and the focus of the displeasure of their constituents, whether individual voters or special interest groups. The later consideration explains why legislators will also want a moral consensus among their constituents. One reason that recent medical law has developed in the courts much more than in the legislatures is that judges are much more sheltered from political pressures than legislators, who must constantly think about their prospects for reelection. A moral consensus would free legislators to regulate the new reproductive technologies when such legislation would be socially valuable.

The context of adjudication introduces rather different considerations. Any careful examination of recent cases in medical law reveals that judicial decisions often hinge upon moral judgments. Now what moral judgments may properly be introduced into courts of law? In the United States, judges have often, and I believe rightly, argued that the courts ought to appeal to the public morality, not to the moral convictions of the presiding judge or judges. There seem to be at least two important reasons to adopt this principle. First, the central function of any court is to settle disputes peacefully. What is being adjudicated in any case, at least in an adversarial legal system such as ours, is some conflict between the parties before the court. Now the moral convictions of the judge will often be personal and will probably not be shared by both parties to the case; often neither party will agree with the moral judgment of the judge. But public morality is a shared morality; where there is nothing like a moral consensus in the society there simply is no public morality. Where a public morality does exist, its moral standards may well be shared by the disputants and, in any event, will be widely accepted among their friends and acquaintances. Hence, these public standards will usually be reasonably effective in settling the dispute that has occasioned the court case.

Second, legal certainty is of great value in any society. The degree of certainty in the law is measured by the degree of confidence with which one can predict how the law should and will be applied to future cases. If the law is uncertain, those subject to it cannot know how to act to conform to that law or to protect themselves from penalties imposed for actions that may subsequently be judged to have been illegal. This danger is a very real concern of medical practitioners aware of the increasing numbers of suits for medical malpractice, the increasing amounts awarded as a result of adverse judgments in such suits, and the rapidly escalating costs of medical malpractice insurance. In addition, their patients and the subjects of clinical research can feel secure in their legal protection only when legal certainty exists. Now if judges appeal to their individual moral judgments in deciding cases before their courts, it will be very difficult to predict just how the law will be applied to individual and variable cases, for this will vary depending upon the personal moral views of the judge or judges who happen to be deciding this or that particular case. If the courts appeal to public morality reflecting a moral consensus within the society, legal certainty will be much greater. There are, then, several reasons why the law, as a whole or in part, will want a moral consensus concerning the uses of the new reproductive technologies.

Can the law contribute in any important way to the moral consensus so valuable to the law? Before we attempt to answer this question, we must define the kind of moral consensus we have in mind. From the papers for this conference, I have gleaned three crucial variables – content, extent, and degree. Our present concern is the morality of technical intervention in human reproduction. My guess is that the best we can hope for is a low-level consensus on specific moral principles concerning such intervention, for example, that experimentation on the human embryo is morally permissible up to the fourteenth day of pregnancy *or* that after the fetus has become viable the mother has a moral obligation to submit to medical treatments necessary for the health of the unborn child, provided these do not impose excessive risk on her. A deeper agreement on fundamental ethical theories from which one might derive these principles appears well beyond the reach of the law and, in any event, unrealistic. And a shallow agreement on each and every particular intervention seems excluded by the variability of special circumstances and the diversity of perspectives from which these instances will be viewed. Presumably, the extent of any consensus resulting from any given legal system would be limited to the jurisdiction of that legal system. Moreover, one can hardly hope for unanimity even within a single society. The best one can hope for is widespread agreement. Even among the majority of citizens, one would not expect the highest degree of agreement. While some may wholeheartedly endorse some set of moral principles concerning technical intervention in human reproduction, others will merely accept them in the far weaker sense of not objecting to them. Our hopes, then, must be modest.

How might we hope to achieve even this modest moral consensus through the law? One way is to argue that any legally valid judicial decision must be grounded upon one or more authoritative legal sources. This legal reasoning by which judges ground their decisions in particular cases might produce a moral consensus. Any judicial decision applies some specific legal rule to the case before the court. If the judicial reasoning could convince the general public that this legal rule reflects some comparable moral rule, the result would be precisely the sort of low-level moral consensus we are seeking.

But how could it achieve this? Legally, the court must ground the rule it applies upon authoritative legal sources. But this could lead to some *moral* conclusion only if these legal sources were morally accepted as reflecting moral rules or principles with a similar content. This does sometimes happen. Citizens often believe that the fundamental principles of their constitution and many, not all, statutes and common law rules have a morally sound content.

There is another way in which the judicial reasoning by which some decision is grounded might possibly achieve a moral consensus. What needs grounding is the application of some rule to the case at issue. Hence, judicial reasoning must pay as much attention to the facts of the case as to the sources of the rule it applies. And often factual information is taken to have legal relevance because of implications concerning the interests of the parties before the court or public welfare or social justice. Now these are morally as well as legally relevant factors. Hence, by revealing factual information that can be recognized by the general public as being morally relevant in a way that supports the applicable rule, the judicial grounding may also achieve a moral consensus on that rule.

This optimal outcome is far from inevitable, however. All that legal validity requires is grounding the judicial decision on legally authoritative sources; these legal grounds need not be morally acceptable to the judge, much less the general public. And the legal relevance of factual information need not imply any moral relevance because positive law is a human creation that could be other than it is and, therefore, need not coincide with morality.

A second approach to moral consensus is through the process of legislation. Legislation, at least in a democratic society, reflects, and is supposed to reflect, a compromise between the diverse preferences and interests of the members of that society. Although the legal validity of any statute does not depend upon its grounding, its enactment will in practice require a compromise achieved in large measure through discussion and argument between the legislators, and more widely within their constituencies. This legislative reasoning might also produce a moral consensus on the matter at issue.

But where does a moral element enter into political compromise? The interests to be integrated into any institutional compromise are not purely self-centered and material; the preferences of those represented by the legislators reflect their moral convictions as well as their more narrowly economic interests. Moreover, to the degree that any compromise really does reflect the preferences and interests of all those affected by it, it can claim social justice. Therefore, there will be some tendency for legislative compromises to reflect and be morally justified by moral considerations.

Equally important, if not more so, is the means by which legislative compromises are achieved. Legislative reasoning can succeed only if political discussion manages to change the minds of a significant number of those opposed to the proposed statute. Moral arguments will feature prominently in any serious political controversy. Hence, a legislatively acceptable compromise can be attained only if some considerable degree of moral agreement can be achieved during the course of the political debate.

Still, compromise is not consensus. It typically involves for each participant some trade-off between what one wants, and may even believe morally required, and alternatives which are even more undesirable and morally repugnant than the outcome accepted as the best attainable. One suspects that only occasionally will the political debate and legislative reasoning leading to an institutional compromise produce any significant moral consensus.

Both judicial and legislative reasoning have the potential to achieve a modest moral consensus. But neither legally valid judicial reasoning nor politically effective legislative reasoning need be morally sound. There is also the trivial point that sound reasoning is not necessarily convincing to an imperfectly rational public. My own conclusion is that it is reasonable to hope, but not to expect, that we can often achieve moral consensus through legal reasoning.

Sometimes, however, we can and do succeed. When legal reasoning does contribute to a moral consensus in the society, this is because it is one small part of a much larger moral discourse. It is through this larger process that Prof. Kurt Bayertz hopes for modest moral consensus. He suggests that rational argumentation, conceived of broadly as an ongoing process of communication in which different moral viewpoints are confronted with one another and in which each is made more understandable through its grounds, can be seen as a permanent process of consensus building. Properly qualified, this seems to me to be the correct approach.

He very wisely adds that we must abandon any expectation of finding or achieving an absolute consensus. But once we give up this goal, we recognize that we can complain of a thorough, universal, and complete disagreement as little as we can concern ourselves with a thorough, universal, and complete agreement. What we find in our society and in others is a patchwork picture of partial moral agreements and disagreements; moreover, some relative consensus exists and may be enhanced on all levels from abstract ethical theories through general moral principals down to concrete cases ([1], pp. 4 f., 13). This seems to me to be a realistic picture, neither the unbounded idealism of the naive optimist nor the excessive scepticism of the hardened pessimist.

It is these many, but partial and fragmentary, agreements we find when we discuss with friends and strangers moral issues, such as those posed by new reproductive technologies, that provide the materials out of which rational argument can hope to achieve a greater, inevitably incomplete, moral consensus. But how? By what magic can rational discussion produce more agreement out of less? No magic is required, only two aspects of moral reasoning.

A rationally justified belief, whether it be scientific or moral, is one that coheres with the body of our accepted beliefs and with our experiences. This traditional coherence theory of justification has recently been adapted by John Rawls for moral principles in general and principles of justice in particular. I will not here trace the development of his varied proposals through his publications; they are familiar to most of us. What I must do is to make it clear that to my mind our hope lies in wide reflective equilibrium. A narrower reflective equilibrium that merely adjusts moral principles to considered moral judgments and *vice versa* often does little more than systemize one's moral prejudices. But when one attempts to fit moral judgments and principles into a wider context of factual beliefs, scientific discoveries and comprehensive moral theories, perhaps even theologies and cosmologies, moral complacency is much harder to maintain. This is all the more true when one attempts to achieve a reflective equilibrium with others. Reasoning, whether scientific or moral, is a social process. Although no rational person will blindly accept the views of others, neither can she ignore their sincere beliefs or reported experiences. In striving for wide reflective equilibrium on a social scale, therefore, one takes the fragments of the colorful picture of partial agreements and disagreements, both moral and nonmoral, as pieces to be put together, perhaps modified, to form a more coherent picture.

The pieces of the puzzle, agreements and disagreements, do not modify themselves. The second aspect of rational justification is challenge and response. Moral discourse is an ongoing process of challenging accepted beliefs and attempting to defend them with grounds or relevant considerations. Any reflective individual will question her own beliefs, especially when they are found to conflict with other beliefs, factual or moral, or with recalcitrant experiences. But it is the challenges from others, typically those who disagree, that provide the greatest stimulus to reexamine one's convictions and either support them with good reasons or modify them in the light of a wider range of relevant considerations. Since challenges are most frequent and most telling precisely where we disagree, and less common in areas of agreement, there is reason to hope that this process of challenge and response among those seeking a reflective equilibrium will reduce disagreements and tend towards consensus.

But am I not myself indulging in comforting but naive optimism? In a small homogeneous society in which traditional values are solidly entrenched, it might be reasonable to expect that the rational discussion of moral problems would arrive at something approaching a social consensus. But as Professor H. Tristram Engelhardt reminds us, we live in large-scale heterogeneous societies where minority opinions and individual dissent render moral unanimity nonexistent and perhaps unattainable ([2], p. 19).

In an international setting I would not dream of denying that men and women come from diverse cultures and that each lives in a highly pluralistic society. The moral disagreements found in such settings mirror faintly the very real and often more radical disagreements between and within our societies. But all this establishes is where moral reasoning must begin, not where the process of reasoning must end – whether in persisting disagreement or in something like a moral consensus. We cannot predict with any great confidence whether moral discourse will arrive at agreement until we know whether our initial disagreements will prove impervious to reasonable challenges. In addition, let us not forget that the moral disagreements that divide us may be balanced or even outweighed by our fragmentary, but equally real, agreements. If we take moral reasoning as a process seriously, we will not imagine that an initial pluralism must inevitably rule out reaching widespread, probably not unanimous, agreement.

Nevertheless, it might seem that disagreements *must* persist because our moral reasoning is prejudiced, literally pre-judged, on the most fundamental level. Engelhardt argues that the disputes concerning the morally permissible uses of reproductive technologies between the religious believer and the secular cosmopolitan do not appear to be resoluble because the disputants bring with them radically different interpretative schemas for calculating the harms and benefits involved ([2], pp. 25, 27 f, 29 f). He need not rest his case on the strong contrast between those who see a transcendent dimension to human reproduction and those who view it as a purely natural phenomenon. The secular moral perspectives of the classical utilitarian and the Kantian are almost equally divergent.

It is certainly true that we do not ask moral questions, confront moral choices or discuss moral issues with an empty mind. Indeed, for a *tabula rasa* there would be no moral problems and no moral issues to debate. Our moral experiences are formed by and our moral judgments informed by our diverse moral perspectives, often radically different from one another. Still, the fact that our radically divergent viewpoints do not necessarily preclude agreement on the sort of low-level specific moral principles needed to guide technical interventions in human reproduction is shown by the frequency of something approaching an overlapping consensus within limited areas of morality. Although it is John Rawls who has recently appealed to the possibility of an overlapping consensus within a pluralistic society, he is not the first to have noted this phenomenon. Scholars familiar with the history of ethics have often remarked upon the fact that moral philosophers tend to converge upon very similar specific moral principles from their radically different, apparently incompatible, theoretical premises. What this suggests to me is that we are better, and wiser, than our most general and abstract theories.

It shows more importantly that we bring more than our basic interpretative schemas to our moral reasoning. I do not wish to disregard or discount the importance of one's *Weltanschauung* in the way in which one interprets one's experiences, lives one's life or attempts to justify one's convictions. Everything Engelhardt has said on this score is true, but it is not the whole truth. If moral reasoning were purely deductive, the radical differences between our basic perspectives and most central theoretical presuppositions would necessitate different conclusions about the morality of technical interventions in human reproduction. But since rational justification is a matter of coherence, wide reflective equilibrium can build on lesser agreements to be found in our scientific knowledge, specific moral judgments about other areas, and our personal experiences. It is important to bear in mind that our interpretative schemas do not apply themselves; we use them to give significance to our experiences and to inform our judgments in conjunction with a wide variety of additional cognitive and emotional factors. Moreover, we can and often do question our divergent interpretative schemas and sometimes adjust them in the light of less exalted, but more solidly grounded, considerations.

My unabashed appeal to a coherence theory of justification will be rejected by many, for it seems to imply that moral reasoning is essentially circular. How, for example, can one rationally decide whether the state ought to forbid some or all uses of *in vitro* fertilization? Presumably, one must know whether the costs in terms of individual liberty and occasional infertility would be greater or less than the benefits in terms of alternative uses of medical resources and increased determinateness of paternal responsibility for child support. But to assess the relative importance of these specific values, one would need to appeal to some general theory of the good. On my view, however, one cannot know which theory of value to accept until one knows what things are good and to what degree. Thus, wide reflective equilibrium requires that everything be adjusted to everything else, which implies that there is no firm starting point and no foundation upon which to build any moral theory or from which to justify any specific moral conclusions. Considerations analogous to these lead Engelhardt to conclude that the question posed is unanswerable by reason ([2], p. 31).

I would reply that there is a starting point and that there is a foundation. The starting point for each individual is whatever opinions and beliefs she does not doubt at the time she is attempting to respond to whatever challenges have been presented to the belief in doubt; the starting point for any set of persons discussing some moral issue is whatever agreements, moral or nonmoral, they find among themselves. To be sure, any undoubted belief or mutual agreement can be challenged. If so, then it must either be set to one side or shown to be justified by an appeal to what remains undoubted. The foundation is experience. Although any experience can and will be interpreted in terms of one's conceptual framework, accepted beliefs and total *Weltanschauung*, there is something beyond challenge in any concrete experience. These very particular experiences, and not our most general and abstract theories, are the ultimate foundations of all rational justification. But they are too few in number, too poor in content, and too ambiguous in meaning to support any significant body of beliefs. Only when supplemented with undoubted, but not indubitable, beliefs ranging from very specific to highly theoretical can they function in rational justification. Within the process of wide reflective equilibrium, however, experiences and beliefs can and do justify many of our moral conclusions and might well result in moral agreement.

But to say that the process of moral reasoning might transform moral disagreements into moral agreements is not to say that anything like a moral consensus will in fact be achieved. At best, and not all my colleagues will grant me even this, all I have shown so far is that a moral consensus concerning technical interventions in human reproduction is not ruled out by the considerations advanced by Professor Engelhardt. But is there any positive reason to believe that the process of reasoning tends towards agreement rather than disagreement? After all, the Socratic questioning of the traditional Athenian moral consensus and the Cartesian doubt of scholastic dogma shows that reasonable challenges can disrupt agreements just as they can disarm disagreements. I must confess that the power of human reason to achieve social consensus cannot be empirically verified. Modern and more recent moral philosophy has been, if anything, less capable of achieving anything like a consensus through its appeal to reason than traditional religions were through revelation and institutional creeds.

Nevertheless, I believe that we are justified in believing that moral reasoning does tend toward agreement and that any remaining disagreements are primarily the result of nonrational or even irrational factors. For one thing, this is a presupposition of the deontological expressions in our moral language. To assert that an action is morally right, wrong, obligatory or impermissible is to make an implicit claim to rationality, and this in turn involves a claim that all normal persons will agree in the light of an indefinite process of challenge and response. For another thing, moral choice poses a problem only if one presupposes that there is a genuine objective distinction between right and wrong, between rationally justified and unjustified choice. And the objectivity of this distinction requires that at the ideal limit of reasoning, all normal persons will draw this distinction in the same way. To abandon the claim to objective rationality is to deprive an important part of our moral language of its significance and to reduce human decisions to psychological events which may be uncomfortable until made but never subject to rational criticism. In other words, to give up the presumption that moral reasoning tends towards agreement is to admit that there is no genuine moral reasoning and are no moral choices about which one should or even could deliberate. A moral philosopher might assert that this is so, but no reasonable person could live by this philosophy.

I insist, therefore, that my optimism is not naive. At the same time, I must confess that it is rather limited. The very basis for my optimism, the multifarious fragmentary agreements from which the process of moral reasoning could fashion something like a consensus, is also a reason to expect that any consensus achieved will fall far short of a complete systematic unanimous agreement. As I have already noted, the best we can reasonably hope for is a widespread fairly weak acceptance of a few low-level moral principles applicable to technical interventions in human reproduction. As a moral philosopher striving to discover and establish the true ethical theory, its derived moral principles and the methods for its proper application to concrete moral problems, I would likc much morc. As a philosophcr of law, howcvcr, I bclicvc that such local social agreements on specific moral issues will usually be sufficient for the purposes of legislation, adjudication, and the law as a whole.

Department of Philosophy
Washington University
St. Louis, Missouri, USA

BIBLIOGRAPHY

1. Bayertz, K.: 1994, 'Introduction: Moral Concensus as a Social and Philosophical Problem', in this volume, pp. 1–15.
2. Engelhardt, H.T., Jr.: 1994, 'Consensus: How Much Can We Hope for? A Conceptual Exploration Illustrated by Recent Debates Regarding the Use of Human Reproductive Technologies', in this volume, pp. 19–40.

ALBERTO BONDOLFI

COMING TO CONSENSUS: AN ETHICAL PROBLEM IN LAW AND POLITICS – ILLUSTRATED BY THE EXAMPLE OF REPRODUCTIVE TECHNOLOGIES*

The necessity of making consensus an object of study in social science research, legal philosophy, and ethics is a specifically democratic achievement and duty at the same time. Pre-democratic forms of society have no need to make an issue of the processes involved in coming to a consensus precisely because these societies hardly ever rest upon a consensus emerging from such a procedure. For all of us, however, the search for a rational strategy of consensus, primarily in ethical questions, poses the only feasible alternative to *bellum omnium in omnes.*[1]

This paper will attempt to explore some aspects of the concept of consensus and then illustrate them by using the example of reproductive technologies. The basic intention of the following reflections is a normative-ethical one, which presupposes that in principle there need not be a basic contradiction in any ethical decision between the notion of a moral consensus and the ideal of *the autonomy of the will.*

I. THE THEOLOGICAL ROOTS OF THE CONCEPT OF CONSENSUS

If one attempts to reconstruct the history of the concept of *consensus* in the history of European philosophy, one must necessarily confront *Theologumena.*[2] *Consentire* is, at least in scholastic literature, circumscribed in such a way as to enable it to subsume the act of faith. As early as Augustine the act of faith was described as a *cum assensu cogitare.*[3] Thomas Aquinas speaks, therefore, of *consensus* as an act of the will, which leads to action. The ability to reach consensus is a specifically

Kurt Bayertz (ed.), The Concept of Moral Consensus, 123–128.

human characteristic, which is structured hermeneutically (as a *consentire interpretative*) and which represents the human capability to reason. In the act of "permitting", therefore, a consensus is at work, in the sense that the act[4] that is being self-willed was first recognized as correct.

Even if in the tradition of scholastic ethics consensus has played an even smaller role, the elements just described nevertheless come again to the fore in contemporary philosophy, in a secularized context. The aim of this paper is to typify such ideas of consensus, in order to determine their indirect ethical function. We shall describe the ideal types of two extreme conceptions of consensus; and, from the impossibility of representing them in a consistent manner, we shall derive the necessity of a "*middle-way*" concept of consensus.

II. OPPOSING CONCEPTS OF CONSENSUS AND THEIR NORMATIVE CONSEQUENCES

Concepts of consensus and theories of truth are closely tied together in a reciprocal relationship. The more truth is perceived as a given entity, the more consensus is defined as an act of the will. On the other hand, an extremely nominalistic conception of truth also favors a perception of consensus as a pure decision of the discerning subject. Such extreme interpretations of consensus exhibit ethical problems. Wherever consensus is seen in close relationship to a *substantial* conception of truth and is perceived as nothing more than passive consent, an ethical act by a free subject can no longer be achieved. And in the sceptical understanding of consensus we have just mentioned – where consensus is looked upon purely as a declaration of one's consent, without any argument whatsoever – the ethical is confused with the factual results or with that which is laid down by authority. Both variants appear to us to contradict an autonomous understanding of the ethical act and, consequently, are not in accordance with what K. Bayertz has called "*moral consensus*" [2].

A *moral consensus* should find its point of orientation in a *middle-way definition*, in which the very act of agreement implies a minimum affirmation of values. Consensus understood in this manner allows, consequently, a common affirmation of the valuable or a common avoidance of the worthless, which can coexist with varied and particular material-ethical options, without again calling into question the established consensus.

This understanding of consensus also presupposes that there is a justified social need to act according to norms only partially affirmed. If such a phenomenon of internalized priorities of action did not exist, common social life would be totally paralyzed. One can hardly imagine, in fact, a community in which social acts become possible only when everyone is in substantial agreement with the totality of all options of action open to political authority.

From this perspective, morally motivated dissent is not only an "achievement", as Bayertz stresses, but it is also the presupposition out of which a morally motivated consensus can emerge. The interlocking of consensus and dissent should, therefore, be experienced as a *moral learning process* which is to be mastered. This "consensus-dissent" is surely more difficult in social practice than reaching decisions by means of a majority vote.

This form of agreement can have value not only for actual norms, but also for the evaluation of empirical facts. Later we shall try to illustrate the significance of a *consensus via negationis* with the example of the "status of the embryo". In the realm of law, with its own tradition, this common exclusion of possible evaluations of empirical data has been known for some time: one need only think of the elimination of the definition of a corpse as *a thing*, although the same legal system would not make a positive definition of the same normative for all its citizens.

III. SOME CONCLUSIONS FOR BIOETHIC CONVERSATIONS

The search for a moral consensus is present in all topics of bioethical research, but it is particularly intense in the area of human reproductive technologies. It is not possible in this context to undertake an analysis of the contents of the arguments presented here for or against the ethical allowance of such techniques. This would be out of the scope of a short and precise response. Our aim is rather to illustrate what a consensus *via negationis* can achieve within the context of this problem.

3.1 Extreme Positions with Respect to Human Reproduction

An extreme position which must be considered radically incapable of consensus would be that in which the problematic issue is looked upon as an exclusively private concern. Without succumbing to a premature moralism, one must nevertheless recognize that reproductive technologies, which have social consequences, must be considered a problem which concerns everyone and which, therefore, needs to be publically and legally standardized.

A second extreme position would consider the issue to be *fundamentally* out of man's hands. At this point those speaking up are primarily religiously motivated groups who believe that human reproduction is *per definitionem* something between the concerned couple and God. As much as this position deserves respect, it nevertheless contradicts the presuppositions of an open society in which conflicts must be solved on the presupposition *etsi deus non daretur*.

Prior to all of these difficulties and differences one should have at least reached a prior consensus, so that future legislation will necessarily comply with the arguments of ethical reflection and, subsequently, not be resolved by dogmatic positions. This would also guarantee that ethics, rather than functioning as carrier of ideologies within the realm of legislation, would act as an authoritative agency of reflection of accepted norms. What does this mean for the extreme positions just mentioned?

As far as the first extreme, "liberal" position is concerned, one can maintain that the "right to biological fertility" can only partially be demanded. Since biological infertility cannot be defined as an "illness", in the fullest sense of the word, an unqualified right to a specific, and only partially effective, method of treatment cannot be deduced. A consideration of the other extreme positions, furthermore, excludes an *absolute prohibition* of reproductive technologies and leads to an ethical affirmation of some rights and obligations.

Doctors have the right and the duty to combat effectively infertility. This, however, should be done within the framework of an overall strategy, in which one not only thinks of fulfilling as much as possible the subjective wishes of childless couples, but seeks rather, above all, through preventive means, to fight the causes of infertility. Only then does one try to fulfill the wishes of the individual couple. This would also solve part of the allocation question in this particular area.

No complete and clear judicial-political strategy has emerged from these reflections and the rejection of extreme positions. What has emerged, however, is an *ethical preference* for those regulations which permit reproductive technologies only under *clearly defined (medical) grounds*. When extreme positions are collectively negated, one always avoids the question of *basic rights* in these areas of life and keeps to the "*middle-way* category" of justified or legitimate concerns.

3.2 Convergence of Opinions in the Question of the Status of the Embryo?

As far as embryo transfer is concerned, we are of the opinion that the ethical problems cropping up in this regard may (must) also be solved without a completely sure decision regarding the *ontological status of the embryo.*[5] Since the question involved is not of an empirical nature, a substantial consensus in this area could not be reached. On the other hand, one must recognize that normative choices are inevitable. In the face of what only seems to be a hopeless situation, should one not also attempt to find the middle way which excludes extreme positions?

Definitions which go to either extreme, that is, which minimize or maximize, can bring the concerned persons and those around them into an unfortunate ethical predicament. If one should maintain, for example (here we mention the minimalist variant), that human life is present only where man is able to carry out independently specifically human acts/actions, there would be the danger, in our opinion, of considering those incapable of functioning effectively as human beings (e.g., the mentally deficient) as no longer worthy of protection. Every definition of this type would have to differentiate clearly between "valuable" life and "worthless" life and a decision tied to this extreme definition would oppress those who are not in a position either to speak about their own existential situation or to declare and defend their own existence as something worthy of the same protection. The minimalist definition leads then to norms which favor the strong. If one proceeds from the other extreme definition, namely, that human life in the embryo stage is always as worthy of protection as the life of an adult person, one likewise ends up in an extremely paradoxical practical situations. Every accepted forfeiture of an embryo would then be equated with a qualified annihilation of a human individual.

Both extreme positions lead, therefore, to a paradoxical dead end. Would it not be better for one to agree upon an assertion about what an embryo *is not*, instead of necessarily seeking to converge on a positive answer to the question. We herewith conclude that an embryo is to be considered neither as a *thing* nor an *individual in the fullest sense of the word.* Such a middle way is not irrelevant as far as norms are concerned, but rather leads, at the very least, to a few basic choices.

Without wishing to claim that an embryo is a complete person, one may nevertheless say that its "*human quality*" (biologically considered) is sufficient to forbid a purely instrumental use of the embryo and to allow only those actions which are necessary to bring an in vitro fertilization effectively to completion. This evaluation does not seem to yield rich results. We prefer, however, a more limited consensus, resulting from reasoning or argumentation, rather than forced or rhetorical consent.

Institute for Social Ethics
The University of Zurich
Switzerland

NOTES

* Translated from German by Doris Wagner-Glenn.

[1] The expression, as is known, goes back to Hobbes ([3], 1,12).

[2] For a reconstruction of the history of the consensus concept, cf. [5].

[3] Cf. ([1], 2,5 p.962). For a general understanding of theological consensus, cf. [4].

[4] Cf. ([7], Q. 23, a.3, in c.).

[5] Cf. in this regard [6].

BIBLIOGRAPHY

1. Augustinus, A.: 1865, 'De praedestinatione sanctorum', *Patrologia Latina* 44, 959–962.
2. Bayertz, K.: 1994, 'The Concept of Moral Consensus. Philosophical Reflections', in this volume, pp. 41–57.
3. Hobbes, Th.: 1959, *Vom Bürger*, F. Meiner, Hamburg.
4. Newman, J.H.: 1962, *Entwurf einer Zustimmungslehre*, Grünewald, Mainz.
5. Oehler, K.: 1961, 'Der Consensus omnium als Kriterium der Wahrheit in der antiken Philosophie und der Patristik', *Antike und Abendland* 10, 103–129.
6. Thévoz, J.M.: 1990, *L'embryon entre nos mains*, Labor et Fides, Genève.
7. Thomas Aquinas : 1964, *Questiones disputatae de Veritate*, Marietti, Torino.

LAURENCE R. TANCREDI

THE EMPIRICAL LIMITS OF CONSENSUS: CAN THEORY AND PRACTICE BE RECONCILED?

The discussion in this volume has focused on the philosophical or theoretical issues concerning moral consensus. That "consensus" has become an important feature of bioethical analysis with hospital ethics committees and national commissions has also been noted along with the recognition of certain benefits to be achieved through this process. According to Engelhardt [12], consensus has the benefit of reducing conflicts among political groups and thereby increasing cooperation for intended goals. Consensus also provides a mechanism whereby diverse community groups and consumers of health care services (such as those who desire embryo transfer and in vitro fertilization procedures) can have their viewpoints considered. In time, a chronicle of moral positions could be created to provide the framework for a "rational" base for future decisions in these matters. However, the problems of consensus as a method for arriving at moral "rightness" are also well described by Bayertz and Engelhardt, as well as Moreno, who have thought deeply about these issues ([3]; [12]; [27]; [28]; see also [40]; [18]). The main presentations have articulated the important conceptual concerns in assessing the relevance of consensus as an approach to ethical decisions in biomedicine.

This comment will focus not on the theoretical issues but on the practical problems of implementing decision-making by consensus, whether unanimous or not. The thrust of this discussion will be that in the everyday world where decisions with "moral" implications are frequently made by existing institutional structures, forces intimately connected with the process of arriving at consensus act to distort, alter and even at times subvert the validity of the process. The conclusion, therefore, of this exploration of the practical problems of consensus would support

Kurt Bayertz (ed.), The Concept of Moral Consensus, 129–141.

Engelhardt's theoretical position that "normative consensus in the sense of unanimity is unattainable ..." and that, therefore, the state should constrict its regulatory roles to that of insuring that "citizens are protected against fraud and other varieties of unconsented-to harm" ([12], p. 20). Essentially Engelhardt would abandon efforts to institute decisions based on "consensus" for regulating reproductive technologies, but instead would maintain existing ethical, social and legal principles to minimize adverse consequences to patients, consumers, and other users of services.

In assessing the role of consensus for arriving at ethical decisions in reproductive technology, for example, it might be useful to examine the effectiveness of consensus in existing models of decision-making. The most obvious from the legal perspective is the functioning of the jury in deciding on accountability for both civil and criminal acts. In the most serious of these criminal acts, such as murder, the requirements of consensus reach unanimity. In other types of jury decisions consensus may be less exacting. Nonetheless, it would still be susceptible to factors that operate to undercut the substantive, if not moral, validity of the conclusions of the group. The analogy between the "jury" process and an ethics committee assessment of reproductive technology issues is a close one. Both are provided with a framework or set of principles upon which to evaluate data regarding the circumstances of a criminal act or a new medical procedure such as a new form of in vitro fertilization. The jury is also important because in many ways it is prototypic of those administrative committees that come to conclusions about the "appropriateness" of actions and thereby shape social policy. In the medical care system, for example, increasingly "committees" are used to make important health care decisions including evaluating the technical appropriateness of medical treatment.

Various studies over the past twenty years have examined factors that affect jury decision-making. Some of these factors go far beyond rules of evidence, the instructions of a judge, or the traditional merits of legal claims. Factors that affect jury decision-making include personality and socio-demographic (sex, race, socio-economic status) characteristics of jury members [10], the potential impact of a jury's decision on the community or on social policy, the group dynamics among leadership positions in the jury, the jury's reaction to the style of presentation of counsel or of specific witnesses, and the personal histories of individual members of the jury, to name a few.

The potential for the jury to be manipulated and to some extent controlled by outside influences is significant. The results may be an effective neutralizing of the unbiased objectivity expected of a jury when it deliberates. Studies have established, for example, that jury decisions may be significantly influenced by a variety of factors including the gender of the defendant. Juries which are composed largely of males are likely to find a female defendant who is attractive innocent in far more cases than if the defendants were unattractive or a male ([8]; see also [15]; [31]; [34]). Furthermore, when an attractive female is found guilty, despite the severity of the offense, she is likely to be assigned a light punishment. The power play created by gender differences on the jury is also very influential in its final decisions. Studies have shown that male jurors, for example, are likely to comment far more often than female jurors on the same jury ([16], pp. 141f.; see also [24]; [25]). Verbal commenting, of course, is very important in establishing the power within the group. It may assure that the power remains in the hands of men who may thereby effectively dominate in the way the decision turns out in any one case.

Other influences include the way the information is presented to the jury [9]. For illustration, a study was conducted where an actor was prompted to play the role of a critical witness for the plaintiff in a case [2]. He acted out his role in two different ways, using the same transcript of material. To one group of observers he appeared quite positive. He styled his facial expressions and general body language to create an upbeat demeanor. To a second group, using the same transcript he appeared with a negative attitude. He appeared disorganized and unsure at times as he made his presentation. He even seemed impolite and annoyed. Of the group that experienced him as "positive," 61% rated him as credible and were most likely to return a verdict in favor of a plaintiff. Of the group where he presented with a negative demeanor, only 33% saw him as credible. Where he was a positive witness, 72% came out in favor of the plaintiff, in contrast to 22% for the plaintiff when he presented with a negative demeanor.

In addition to the impact of the demeanor of the witness on the jury, the way questions are phrased or information imparted by counsel also influences jury decisions. For example, cross examination of a witness through questions which allow the attorney to create certain images, conjectures or implications may consciously or unconsciously affect the listener's attitude and capacity to objectively evaluate what the witness says. The lawyer engaging in cross examination can carefully choose his words to highlight, obscure, or alter the testimony and affect the listener's interpretation of the witness' response to questions ([9], pp. 1375–1385).

The impact of how information is presented and its influence on an individual's or jury's understanding is of no surprise to scholars in medical ethics who have been concerned with the effectiveness and validity of informed consent. The manner of communication of information to patients and experimental subjects has been shown to influence their ability to make "reasoned" decisions. An important study conducted at Harvard addressed this issue in a sophisticated way [26]. The subjects of this study included a group of ambulatory patients, Stanford University business students and radiologists who were attending a post graduate course at Harvard. In this study the subjects were divided into two groups and were asked to select a treatment they would prefer, were they to suffer from lung cancer. They were given two treatment options, radiation therapy or surgery. The differences in survival between the two treatments are significant at two stages – survival from application of the procedure and at five years follow-up. For example, a 60-year-old patient undergoing surgery of the lung would have an average mortality rate of about 10%. In contrast, radiation therapy for that same individual would have no mortality from the treatment. However, the five-year survival shifts in favor of surgery. For surgery, the five-year survival is as high as 34%, whereas for radiation therapy it is only 22%. Whether the survival or the mortality information was emphasized made a difference in people's choices of the two treatments. If the outcomes were framed in terms of probability of survival, surgery was more attractive than radiation therapy. However, when the emphasis was on the probability of death from the treatment itself, then radiation therapy was more likely to be chosen than surgery. The study concluded that there is no "point in devising methods for the elicitation of patient preferences since they are so susceptible to the way the data is presented, to implicit suggestions, and to other biases" [26].

In addition to the way information is communicated, metaphors and metononyms can be used to selectively influence decisions. This impact is by no means new. George Lakoff and Mark Johnson in *Metaphors We Live By*, an important work published in the early 1980s, demonstrate how metaphors work in a wide range of situations to provide coherence, emphasis and persuasiveness to arguments [22]. In the context of jury decision-making there are many illustrations where certain words, phrases and images affect the perception of jurors about the factual events involved in a legal decision [4]. In one study the subjects watched a film of a collision of two automobiles and were asked to determine how fast the cars were going prior to impact [23]. One group was asked how fast the cars were going before they "smashed,": the second was asked how fast they were going before they "hit". The first group indicated over 40 miles per hour, the second under 35 miles per hour. The difference in the words "smashed" and "hit" were critical to the perceptions of speed by those observing the collision. A week later the subjects were re-studied to assess their memory of the event, in particular whether or not there was broken glass at the scene of the collision. 32% of those who were asked the initial question with the term "smashed" claimed there was broken glass at the scene of the accident. Where the word "hit" was used in the initial study, only 14% claimed that glass was broken at the scene [23].

The leading question is one of the most powerful tools used in cross examination. When a lawyer asks a leading question which carries with it an assumption or implication of knowledge, generally jurors will credit that implication with some truth value, such as that a basis must exist to support the specific premise being suggested [37]. Jurors will often treat these statements as though they were factual. Furthermore, studies have shown that people will frequently remember a message, particularly its content, without remembering the source of that message [37]. Leading and suggestive questions can clearly mislead a jury. They can also, as we have seen in research involving informed consent, mislead those who are asked to respond to the questions. For example, a study was conducted in 1981 of 50 newly admitted patients to a psychiatric unit [1]. The purpose of the research was to evaluate the competency of newly admitted patients to consent to psychiatric hospitalization. The researchers, employing a range of definitions of competency, tested the patients after their admission and concluded that the majority were severely impaired and incompetent.

Fifteen questions were asked of the patients on subjects such as their awareness of the nature of hospitalization, their understanding of the reason admission was recommended, their appreciation of the nature of their condition, their ability to cooperate with the treatment planned, and their awareness of their rights. Some of the questions were clearly leading, containing within them strong suggestions that would lead the respondent and influence those observing the questioning. For example, questions such as the following were most compelling and conclusory: Do you think you need some kind of treatment for your problems? Why do you think the doctor you saw recommended that you come into the hospital? Do you think that you need to be in the hospital to get that treatment?

In evaluating the results, the researchers indicated that nearly half of those admitted did not think they needed hospitalization, and that "only 46% could clearly acknowledge that they had psychiatric problems." This the researchers concluded indicated that the patients were "not engaged in the rational manipulation of information that is a desirable element in any definition of competency". The questions were clearly ideologically loaded and framed so that a negative response suggested that the patient could not rationally manipulate information ([38]; see also [20]).

The ability of jurors to evaluate truth and deception has also been shown through studies to be quite limited, even though much of their evaluation depends on their ability to make such assessments. But studies have shown that frequently there is a mismatch between what a witness is communicating in nonverbal behavior and how it is perceived by others, such as a jury [41]. Frequently, clear evidences of nonverbal behavior associated with deception either go undetected or are misinterpreted by perceivers. Often people will focus on changes in the expressions on a speaker's face rather than scrutinize kinesic or other cues. This is paradoxical since facial expressions are more likely to be under the conscious and perhaps deceptive control of the speaker then are other bodily movements [41].

In addition to factors associated with persuasion, image, and deception, there are other considerations that influence the way juries and comparable committees come out in various decisions. Powerful psychodynamic features of group process may shift the power for decisions to specific individuals on committees. However, shifts in power may occur for reasons unrelated to personality characteristics. For example, studies have shown that juries will frequently select the foreman who will represent the jury, especially in the presentation of its conclusions, based on who is seated at either end of the table when the selection occurs. Hence, the foreman of the jury, who is often perceived as the leader or principal spokesperson, may achieve this status simply by virtue of his/her location at the time the selection is made. Once selected, the foreman may play a critical role in directing the jury to a verdict [36]. In studies of a variety of juries it has been shown that the foreman speaks far more often than the average juror. He or she has essentially been empowered by that selection process to a position of significant influence over the jury ([16], p. 28).

Other factors besides where the individual is sitting when the foreman is selected determine the power balance among members of a committee. These factors include educational background, life experiences, differing intellectual capacities, the persuasiveness of individual members on a committee, and heuristic processes which operate often unconsciously, certainly in subtle ways, to affect not only who is perceived to be the major spokesman for the committee, but also the verdict that the jury will finally accept [5]. The fact that heuristic mechanisms operate to color the way information is processed and decisions reasoned is a very important consideration in assessing the practicality of using consensus to arrive at ethical positions. Tversky and Kahneman [39] and others ([35]; see also [14]) studying the effects of uncertainty on decision-making have revealed unconscious processes involving the past experience of the decision-maker which affect the ability of even experts to rationally assess probabilities of new situations. The three most influential of these are representativeness, availability, and anchoring. These processes involve images, notions, and pre-conceptions from the individual's past experiences which influence attitude and distort the importance of facts involving a decision.

An interesting paradigm of heuristic processes (particularly representation) affecting jury decisions has been proposed by some cognitive psychologists. This has been referred to as the "hindsight bias" [7]. Fischhoff has called this process a type of "creeping determinism" [13]. It works as follows: a group of subjects was provided with a "scenario" that has several possibile outcomes. The subjects were in fact told of one such outcome that actually occurred. They were then compared with a control group who had been provided with the "scenario" but not told of an outcome. The subjects and the control group were asked to assess the likelihood that various outcomes would occur. The subjects overestimated the probability that the outcome they had been told would occur. Even attempts to get these subjects to ignore the outcome they had been told were unsuccessful.

Essentially, the "hindsight bias" involves the subjects' inculcation of one outcome into their understanding of probabilities so that the particular outcome overshadows their judgments about the likelihood of various outcomes. One could see the relevance of this in the court room where information is presented to a jury that they are then told to ignore when considering the legal issue before them. The information cannot be ignored, and pending its content will likely influence their understanding of the probabilities of events and the legal issue before them.

Unconscious or heuristic influences on individual decision-making are powerful in themselves for affecting how we come out on practical issues. These influences are augmented by the dynamics of committees that are determinative of shifts of power to specific individuals who act not only as spokesmen for the committee, but as dominant forces in shaping the perceptions and conclusions of the group.

In the medical care field, a potentially dynamic consensus process has been emerging as a system of accountability of health care services. This system is the "peer review" method for judging the appropriateness and quality of medical services. Studies have shown, however, that there is a wide range of responses by experts on the quality of services provided in any one setting. Discrepancies in the judgments of interdisciplinary groups on quality of clinical care have called into question the validity and reliability of peer judgments. One of the early studies on quality of services pointed out that not only is there a range of individual responses regarding flexibility or harshness of judgments, but even within the same specialty there are differences of focus among the judges on aspects of the clinical care that is provided [33].

One approach to shoring up the validity of peer review has been to set explicit standards or criteria for those participating in the process ([17]; [21]). Early studies have shown that peer review by committee is best achieved if the individual decision-making is structured by pre-established criteria or standards. These standards provide sufficient comparability in the framework of the thinking of those participating in the process that they limit, to some extent, the degree of variability. The relevance of this to consensus on the ethics of reproductive technologies is the benefit of shoring up the process with well defined principles of ethics or a carefully articulated framework for resolution of conflicts.

Even so, the experience thus far with peer review and the jury system reveals the opportunities for manipulation of the system, coopting of individual sensitivities, and strong forces for regression to the mean in the decision-making of the group. Peer review requires a considerable amount of consensus not only for the examination of individual cases, but for the development of criteria for use in prescinding various levels of quality of care. Conceptual difficulties experienced by peer review would likely have their analogues in the use of consensus in ethical decision making [32].

In addition to these factors impacting on the way committees operate in making decisions, there are two additional issues that have to be considered. First, frequently committees are regulatory and structured to respond to issues affecting a particular industry or set of social problems. These regulatory bodies, however, as shown by Roger Noll in his review of the Asch Committee Report of the effectiveness of the Security Exchange Commission (SEC), frequently serve the ends of the institution that they are ostensibly regulating [29]. Regulatory bodies must rely on information about the activities that they are regulating from those who are being regulated as they are frequently the experts in their area. For example, the SEC, over time, developed a close relationship with brokers and other experts of the stock exchange system. Inevitably the SEC took on many of the values inherent to the operations they regulate.

Essentially, information and expertise become the basis for shifting the balance of decision-making prerogatives away from committees that are set up to exert control over those operations [30]. The somewhat symbiotic association, therefore, between a regulatory body and the object of its regulations is complex and often transforms the regulatory body as well as those regulated. Information inevitably flows between the two groups and values become intermixed such that the control body is in some respects subtended by the objectives of the object of control. This phenomenon has been seen in a variety of settings, including the mental health field where lawyers in various states, such as New York, have been hired to insure the rights of patients in the mental health system. They frequently become coopted by the system and incorporate the values of the dominant social institution, often at the expense of patients who are brought into the system.

The second set of conflicts that affects regulatory systems such as consensus committees involves social and economic forces that "pull" committees to decisions as quickly as possible. Exigencies such as administrative and economic factors involved in setting up juries and creating an efficient court system act to discourage indecision on the part of a jury. The pull, therefore, is stimulated by efficiency requirements.

It reflects an actuarial mind set that promotes the smooth operation of the jury to effectively address the tasks of judicial decision-making. In the area of ethical consensus on the use of reproductive technologies, there will be a strong pull in the direction of arriving at timely decisions to meet institutional needs. This requirement will not only affect the process of approving ethical positions, but by virtue of the values underlying efficiency and actuarial objectives, it will affect the substance of those decisions. Even though a jury rarely has a unanimous first vote on a legal issue, studies have shown that the finally agreed upon verdict is generally the decision held by the majority of the jurors in the initial voting ([19]; see also [11]). The "pull" is toward maximum conformity.

In the jury situation, decisions or verdicts are arrived at through informational processes as well as normative ones. The normative influences include responses to social pressures that involve administrative simplicity, efficiency and concern about the down-stream effects of decisions. These pressures will also operate perhaps in a different form to affect the kinds of decisions that are arrived at by consensus committees in the field of reproductive ethics ([20]; see also [6]).

CONCLUSION

This comment has focused on the dissonance between theory and practice in the use of the consensus process to make social decisions. The practical applications of decisions by committee have revealed a host of factors, internal and external to the constitution of the committee, that influence not only the process of decision-making, but the substance as well. Much of the research on these issues has involved the jury system, which serves as a close analogy to a consensus process concerned with the ethics of reproductive technologies.

As we have discussed, the jury system involves individuals who are shaped by their histories, by factors of group dynamics, and also by the normative social values that impact on all of us at various stages in our lives. Individual members of juries must work to assess information on civil and criminal members and relate their individual views to other members of the jury in order to arrive at a decision or verdict for the court. Dynamic processes such as regression to the mean, potential manipulation of decisions through the manner of presentation of information, power balances among personalities within the group, influences of heuristic processes, the tendency of regulatory bodies to serve institutional ends, and the general values of efficiency and efficacy that frequently propel committees to meet desirable societal requirements may act to preclude an open, honest and ethical examination of issues by committees. The results can be jury decisions that are not based on rational evaluations using principles of justice.

Where ethical assessments are involved, the consensus process offers opportunity for distortions of important principles such as fairness, equity, and personal autonomy to name a few. The distortions are potentially more serious because they result in decisions that may be institutionalized to influence similar activities in society. By virtue of its reliance on the "voice" of many producing a unitary position, the danger of empowerment of the views expressed is considerably enhanced. It is the combination of factors – the potential influences on the rational evaluation of information and the unusual empowerment created by ethical decisions of committees – that the argument can be strongly posed that ethical issues in reproductive technologies should not be handled through committee decisions.

Theory and practice of ethics by consensus may not be reconcilable. This would suggest that the ethicist would be more useful outside of a consensus process. He or she should perhaps see his/her role as a critic of the process of social and medical decisions as they unfold. By remaining outside of the system, the ethicist can maintain a clear unbiased view, one not blinded by the closeness and conflation of institutional and committee objectives that inevitably affects regulatory efforts. It is for this reason that the pragmatic concerns of ethics by consensus would support the Engelhardt position of limiting the purely regulatory role to that of insuring that participants in reproductive innovations are protected against fraud or other types of "unconsented" harm.

University of Texas
Health Science Center at Houston
Houston, Texas, USA

BIBLIOGRAPHY

1. Applebaum, P. and Roth, L.H.: 1981, 'Clinical Issues in the Assessment of Competency', *American Journal of Psychiatry* 138, 1462–1467.

2. Ashton, R.H. and Ashton, A.H.: 1990, 'Evidence-Responsiveness in Profession Judgment – Effects of Positive Versus Negative Evidence and Presentation Mode', *Organizational Behavior and Human Decision Processes* 46, 1–19.
3. Bayertz, K.: 1994, 'The Concept of Moral Consensus', Philosophical Reflections in this volume, pp. 41–57.
4. Bell, B.E. and Loftus, E.F.: 1989, 'Trivial Persuasion in the Courtroom : The Power of (a Few) Minor Details', *Journal of Personality and Social Psychology* 56, 669–679.
5. Bell, D.E.; Raiffa, H. and Tversky, A. (eds.): 1988, *Decision Making: Descriptive, Normative, and Prescriptive Interactions*, Cambridge University Press, Massachusetts.
6. Caplan, C. and Miller, B.: 1987, 'Group Decision Making and Normative Versus Informational Influence: Effects of Type of Issue and Assigned Decision Rule', *Journal of Personality and Social Psychology* 53, 306.
7. Casper, J.D. *et al.*: 1989, 'Juror Decision Making, Attitudes, and the Hindsight Bias', *Law and Human Behavior* 13, 291–310.
8. Chaiken, S.: 1979, 'Communicator, Physical Attractiveness and Persuasion', *Journal of Personality and Social Psychology* 37, 1387–1397.
9. Conley, J.M. *et al.*: 1978, 'The Power of Language: Presentational Style in the Courtroom', *Duke Law Journal* 6, 1375–1399.
10. Dane, F.C.: 1982, 'Effects of Defendant's and Victim's Characteristics on Juror's Verdicts', in N.L. Kerr and R.M. Bray (eds.), *The Psychology of the Courtroom*, Academic Press, New York, pp. 83–115.
11. Davis, R.: 1980, 'Group Decision and Procedural Justice', in M. Fishbein (ed.), *Progress in Social Psychology*, Volume I, L. Earlbaum Associates Incorp., Hillsdale, N.J.
12. Engelhardt, H.T. Jr.: 1994, 'Consensus: How Much Can We Hope for? A Conceptual Exploration Illustrated by Recent Debates Regarding the Use of Human Reproductive Technologies', in this volume, pp. 19–40.
13. Fischhoff, B.: 1975, 'Hindsight/Foresight: The Effect of Outcome Knowledge on Judgment Under Uncertainty', *Journal of Experimental Psychology: Human Perception and Performance* 15, 190–194.
14. Fischhoff, B. *et al.*: 1978, 'How Safe is Safe Enough? A Psychometric Study of Attitudes Towards Technological Risks and Benefits', *Policy Science* 9, 127–152.
15. Greene, E. *et al.*: 1989, 'The Impact of Physical Attractiveness, Gender, and Teaching Philosophy on Teacher Evaluations', *Journal of Educational Research* 82, 172–177.
16. Hastie,R. *et al.*: 1983, *Inside the Jury*, Harvard University Press, Cambridge, Massachusetts.
17. Hulka, B.S. *et al.*: 1979, 'Peer Review in Ambulatory Care: Use of Explicit Criteria and Implicit Judgments', *Medical Care* (Sup) 17 (3), 1–73.
18. Jennings, B.: 1991, 'Possibilities of Consensus: Toward Democratic Moral Discourse', *The Journal of Medicine and Philosophy* 16, 447–463.
19. Kalven, H. and Zeisel, H.: 1986, *The American Jury*, University of Chicago Press, Chicago.
20. Kassin, S.M.: 1990, 'The American Jury: Handicapped in the Pursuit of Justice', *Ohio State Law Journal* 51, 687.

21. Koran, L.M.: 1975, 'The Reliability of Clinical Methods, Data and Judgments', *New England Journal of Medicine* 293, Part 1, 642–646.
22. Lakoff, G. and Johnson, M.: 1980, *Metaphors We Live By*, University of Chicago Press, Chicago & London.
23. Loftus, E. and Palmer, J.: 1974, 'Reconstruction of Automobile Destruction: An Example of the Interaction Between Language and Memory', *Journal of Verbal Learning and Verbal Behavior* 13, 585–589.
24. Mabry, E.A.: 1989, 'Some Theoretical Implications of Female and Male Interaction in Unstructured Small-Groups', *Small Group Behavior* 20, 536–550.
25. Marder, N.: 1987, Note: Gender Dynamics and Jury Deliberations, *Yale Law Review* 96, 593.
26. McNeil, B.J. *et al.*: 1982, 'On the Elicitation of Preferences for Alternative Therapies', *New England Journal of Medicine* 27; 306 (21), 1259–1262.
27. Moreno, J.D.: 1991, 'Consensus, Contracts and Committees', *The Journal of Medicine and Philosophy* 16, 393–408.
28. Moreno, J.D.: 1994, 'Consensus by Committee: Philosophical and Social Aspects of Ethics Committees', in this volume, pp. 145–162.
29. Noll, R.G.: 1971, *Reforming Regulation*, Brookings Institution, Washington, D.C.
30. Noll, R.G. (ed.): 1985, *Regulatory Policy and the Social Sciences*, University of California Press, Berkeley.
31. Pallak, S.R.: 1983, 'Salience of a Communicator's Physical Attractiveness and Persuasion: A Heuristic Versus Systematic Processing Interpretation', *Soc. Cognition* 2, 158–170.
32. Payne, B.C.: 1977, 'Research in Quality Assessment and Utilization Review in Hospital and Ambulatory Settings', in P.Y. Ertel and M.G. Aldridge (eds.), *Medical Peer Review Theory and Practice*, C.V. Mosby Co., St. Louis, pp. 335–355.
33. Richardson, S.M.: 1972, 'Peer Review of Medical Care', *Medical Care* 10, 29–39.
34. Sigall, H. and Ostrove, N.: 1975, 'Beautiful but Dangerous: Effects of Offender Attractiveness and Nature of the Crime on Juridic Judgment', *Journal of Personality and Social Psychology* 31, 410–414.
35. Slovic, P. *et al.*: 1979, 'Rating and Risks', *Environment* 21, 14–39.
36. Somer, R.: 1961, 'Leadership in Group Geography', *Sociometry* 24, 99–110.
37. Swann, W.B. *et al.*: 1982, 'Where Leading Questions Can Lead: The Power of Conjecture in Social Interaction', *Journal of Personality and Social Psychology* 42, 1025–1035.
38. Tancredi, L. and Weisstub, D.N.: 1986, 'The Ideology of Epidemiological Discourse in Law and Psychiatry: Ethical Implications' in L. Tancredi (ed.), *Ethical Issues in Epidemiological Research*, Rutgers University Press, New Brunswick, pp. 157–181.
39. Tversky, A. and Kahneman, D.: 1974, 'Judgment Under Uncertainty: Heuristic and Biases', *Science* 185, 1124–1131.
40. Veatch, R.M.: 1991, 'Consensus of Expertise: The Role of Consensus of Experts in Formulating Public Policy and Estimating Facts', *The Journal of Medicine and Philosophy* 16, 427–446.
41. Zuckerman, M. *et al.*: 1981, 'Verbal and Non-Verbal Communication of Deception', *Advances in Experimental Social Psychology* 14, 1–59.

PART THREE

MICROINSTITUTIONS OF CONSENSUS-FORMATION

JONATHAN D. MORENO

CONSENSUS BY COMMITTEE: PHILOSOPHICAL AND SOCIAL ASPECTS OF ETHICS COMMITTEES

I. INTRODUCTION

Until recently, the literature of biomedical ethics has not attended in analytical detail to the nature of consensus.[1] This is curious if one considers that the origins of modern bioethics are closely tied to the episodes of apparent consensus about the use of human subjects among scientific investigators, uses that are now regarded as wholly unacceptable. Rather, it has been the concept of autonomy that attracted most attention in bioethics, regarded perhaps as the antidote to the moral pitfalls of consensus.

The possibility that consensus is nevertheless an important subject for bioethical analysis has emerged partly due to the appearance of panels such as national ethics commissions and hospital ethics committees. These groups usually operate on the basis of consensus, raising the question whether there are better and worse instances of consensus. Alternatively, perhaps participants on these panels are subject to an elaborate form of self-deception, and their consensus is always as morally suspect as any other.

After explaining how consensus decision-making relates to the development of policies to guide the use of reproductive technologies, I analyze the idea of consensus from historic and conceptual standpoints. Next I show how social psychological research can illuminate and improve the processes of small group decision-making, taking the ethics committee as an example. Then I consider the implications of institutionalized consensus decision-making for the field of bioethics. Contrasting the "dynamic" consensus that emerges from group interaction

Kurt Bayertz (ed.), The Concept of Moral Consensus, 145–162.

with the "static" consensus exemplified by public opinion polls, I suggest that the hope for moral authority in consensus decision-making lies in the connection between the dynamic process and its result. Finally, I conclude with some remarks on the significance of the dynamic/static distinction for any future consensus on policies to guide the uses of new reproductive technologies.

In this paper my concern is with the nature of consensus from a secular bioethical standpoint. I believe that the attempt to create a secular biomedical ethics is precisely what has made the field so challenging. This is not to say that ideas and metaphors to be found in religious traditions may not be immensely stimulating and useful in secular bioethics; indeed, we could hardly do without them. But when one invokes a religious bioethics, consensus has already been presupposed, at least to some substantial degree. No such presumption applies in the case of the secular bioethics that has emerged in our time.

II. DEVELOPING CONSENSUS ON THE NEW REPRODUCTIVE TECHNOLOGIES

Consensus on the ethical use of the new reproductive technologies (including in vitro fertilization, embryo transfer, and third-party assisted reproductive techniques), may express itself through institutions (e.g., clinics and hospitals), professions (e.g., obstetricians and genetics counselors), and political entities (e.g., provincial governments, national governments, and the United Nations Organization). Ethics committees are an increasingly popular form of decision-making in institutions, especially in the United States. Professional groups may also create ethics committees to monitor ongoing developments, or they may create commissions on a more or less *ad hoc* basis to recommend far-reaching policies. The commission model for assessing new bioethical questions is a format that many provincial and national governments have already utilized [22]. Although I am mostly concerned here with the institutional ethics committee, it will be seen that to a great extent the philosophical problems concerning moral consensus are identical for committees and commissions.

Ethics committees may consider questions raised by the reproductive technologies either on a case-by-case basis or in the course of developing recommendations for policies to govern the use of those technologies in the institution. For example, an ethics committee may be asked to help resolve a disagreement concerning the use of some technique in a particular case (such as a couple's request for the implantation of fertilized ova in a surrogate), or it may be asked to help develop an overall policy to guide the institution. Of course, ethics committees are mainly identified with concerns about decisions to terminate life-sustaining treatment. The following discussion about consensus in ethics committees is as relevant to those problems as it is to controversies about the use of reproductive technologies.

III. HISTORIC AND CONCEPTUAL VIEWS OF CONSENSUS

Partridge has called consent and consensus "persistent but elusive" ideas in intellectual history; both are fundamental but difficult to characterize. Classical authorities like Plato and Aristotle did not regard mere consent of the governed as a sufficiently powerful explanatory device to account for the origin of civil life. Rather, biologic needs for food and shelter supposedly prompted the first moment of social organization, giving civil society a virtually organic cast. In the sixteenth and seventeenth centuries, when rulers insisted on their "divine right", the underdeveloped notion of consent of the governed was sharpened into its recognizably modern form that sees each individual as a rightsholder.

As distinguished from consent, consensus appears to have reemerged in the modern period as once more part of an organismic account of the ongoing cohesiveness of civil society, following the theoretical consent to a social contract. Thus J.S. Mill followed Comte in defining "consensus" as the mutual influence of every part of society on every other part. If political theorists are mostly identified with the idea of consent, consensus has mainly been the turf of sociologists interested in social integration and stability. In turn, consensus has come to have a more restricted meaning in recent sociology than its earlier global significance in an organic model of society. Many sociologists emphasize viewing consensus in terms of voluntary agreement with certain objects, such as norms or values, that can harmonize the behavior of many individuals. This widespread concurrence is often embodied in custom or tradition.

Let us turn now in a more focussed way to the recent discussions of *moral* consensus specifically, from sociological, political, and philosophical standpoints. In the Durkheim-Parsons view consensus is a requirement for social stability, for without it there could be no agreement about what counts as the "good society". This claim must be qualified in at least two ways. First, it is obvious that there are many senses of "stability", and some areas of stability are more important than others. For this formulation to be useful, those areas must be distinguished. Second, the moral consensus may of necessity include the determination by different social groups not to push their local vision of the good society in all its details too hard, lest they create conflict with others. Consensus about the good society, if this is a sensible idea, appears to include an impetus to find procedural accommodations.

The sociological standpoint is mainly concerned with describing the various ways that consensus can be used to encourage a sense of social unity. Thus from a functional point of view, a social system may require more or less explicit social agreement, depending on the circumstances. For example, in those cases where a question is regarded as trivial a group will signal this by acquiescence to the *status quo* or to direction given by authority figures. At the other extreme, potentially controversial questions may also be decided by acquiescence if a society gives a sufficiently high value to the preservation of cohesion. Matters which are either not likely to engender social conflict or which are regarded as too important to be left to mere passive agreement could be submitted to a "head count", referred to in parliamentary terms as "the sense of the house", so that members can be sure that roughly most of their fellows concur. Finally, explicit voting can sometimes reveal a consensus, though most often it is avoided if it is desirable to avoid showing clear rifts in social agreement [18].

The political standpoint is mainly concerned with the several functional forms of consensus in authorizing government action [9]. Thus in "permissive consensus" the public is prepared for policies that the government will later be enabled to pursue. Expert panels are frequently convened by government bodies to help prepare the public for some new idea or program [12]. "Supportive consensus" refers to a practice by competing groups within a government – e.g., rival political parties – to continue some settled policies regardless of which group is currently in power. Notice that this form of consensus may have little or nothing to do with democratic processes, for it relates to cooperation among established rival powers rather than to consent of the governed. In "decisive consensus", the polity authorizes the government to take some particular step, though again the government may not be one that on the whole operates through consent: an example would be a people's decision to authorize a despot's defense of the homeland from a rapacious foreign invader [9].

The philosophical standpoint is mainly concerned with differentiating the possible objects of consensus. In general, there may be procedural consensus or substantive consensus. Procedural consensus is operative when there is agreement about the rules and/or methods that will be followed in resolving actual or possible conflicts about substantive matters. In turn, substantive consensus is agreement with one of a number of alternative and conflicting points of view. In practice, there is logical "slack" running in both directions between procedural and substantive consensus: from the fact that there is one sort of consensus the other may not be inferred.

Within substantive consensus one may further distinguish between consensus about particular questions and consensus about principles, or what I have elsewhere called "deep consensus" [11]. A useful example is Stephen Toulmin's report of his experiences as a member of a blue-ribbon national commission setting standards for the use of human subjects in research. Toulmin found that the commissioners had far less difficulty reaching agreement on specific policies than was the case if each attempted to identify his or her own moral reasoning behind the common conclusion ([19], pp. 270–71). Toulmin's observations are surely even more relevant to society at large. In his recent work Rawls has invoked the idea of "overlapping consensus", a spatial metaphor intended to suggest that the members of a pluralistic society will agree on some ideas and values but will not all agree on the same ones. To Rawls this suggests that it is often prudent to refrain from attempts to gain social agreement on certain intractable controversies [17].

The situation is surely complicated if one sees the overlap in its horizontal as well as vertical dimensions. That is, the roughly overlapping beliefs that imperfectly hold the members of a pluralistic society together could be "shallow", in the sense that there is general agreement about particular cases, or "deep", in the sense that there is general agreement about principles, or deeper still, in the sense that there is general agreement about theory. These possibilities are currently being discussed by authors working on methodological problems in bioethics. In the next section I will relate this discussion to the idea of substantive consensus. It should be understood that the following exploration now refers specifically to moral consensus unless otherwise stated.

IV. SUBSTANTIVE CONSENSUS: PRINCIPLES, CASES, OR THEORY?

How deep can moral consensus hope to go? Some of the most interesting recent work on bioethical method has important implications for this question. If there has been a "received view" about bioethical methodology since the early 1970s, it has surely been that from various moral theories one can infer a discrete set of principles such as autonomy, beneficence, and justice, and these "mid-level" principles can in turn be applied to particular cases [2]. In terms of the above discussion of levels of consensus, even if some relevant decision makers have identical preferences concerning the desired outcome of a particular case, they might well discover that, in spite of arriving at a common preference, they applied the principles a bit differently; or if in agreement at these levels, they might well discover differences in the moral theory or theories they select to ground the principles.

In practice these bioethical principles are not usually themselves subject to much attention, save among the ideal theorists. Rather, the principles are normally offered to clinicians and students as "guidelines" that can help to give order to a morally disorienting state of affairs. The principles are therefore the beneficiaries of a presumed consensus, though often as discussion proceeds there is the realization that significantly different weighting of the principles may follow from different moral theories. Still, anyone who has ever worked through an actual clinical case with a group of anxious health care providers appreciates the availability of the bioethical "mantra": autonomy, beneficence, and justice.

The regnant emphasis on principles in bioethics has come in for increasing criticism lately. The criticism has come from two directions: those who advocate a case-based or "casuistic" approach [8], and those who advocate a unified theoretical approach [3]. In general, these critics of "principlism" in bioethics argue that the mere recitation of principles divorced from actual cases and not framed by a moral theory is on the one hand arid and on the other hand groundless. As a criticism of principlism in bioethics, a suitable paraphrase of Kant might run: principles without cases are empty, principles without theory are blind.

Thus on one side the proponents of a modern casuistry urge that the moral life in its richness can only emerge from a wide variety of actual situations. Rather than an ethical system that applies principles to cases "from the top down", they argue that in practice principles emerge through experience with real cases, or "from the bottom up". There are various techniques in the art of casuistry, but vigorous use of analogy is among the most important. For example, by starting with paradigmatic exemplifications of certain common precepts or maxims (viz., "Don't kick a good man when he's down"), the elements of the paradigm case can be systematically varied until one discerns which are its essential features, without which a case would fail to satisfy the maxim.

For those interested in the development of moral consensus the casuistic approach has its attractions. An ethics committee could reflect upon a number of historic cases, both from its own institution and from others, and array them on a continuum according to the degree to which their outcomes satisfied some particular maxim. When considering future cases the members would be prepared to identify the crucial elements that distinguish them from the paradigm cases. Consensus on the case at hand should flow from this process.

In assessing the "new casuistry" in bioethics, Arras notes the ambiguous role of theory in it. Sometimes the new casuists seem prepared to latch onto any source of moral guidance that happens to be lying around, while at other times they seem to give pride of place to moral theories devised by academic philosophers. Appealing as the notion of a "case-by-case" ethics might be, particularly when one wants to avoid higher-level philosophical contention, Arras concludes that casuistry at most yields a theory-poor rather than a theory-neutral method [1].

Insistence upon consensus on a particular moral theory has been far-and-away the least popular option in secular bioethics. In fact, it might not be hyperbolic to assert that one of the features that has distinguished moral philosophy and "applied ethics" has been the latter's rejection of hegemony by a single theory. After all, it was the very effort to develop an analytic or "metaethics" to adjudicate among the moral theories that led finally to the well-known aridity of Anglo-American moral philosophy by the 1960s. These conditions finally gave rise to applied ethics and to a renewed interest in normative questions. Moreover, the idea that a single moral theory can be applied, in deductive-nomological fashion, to specific practical ethical problems, is commonly derided in bioethics as an "engineering" approach. Yet Clouser and Gert call attention to the shortcomings of a bioethics, whether based on potentially conflicting principles or on intrinsically inconclusive cases, that is theory-poor. Perhaps the advantages of consensus on a single, unified ethical theory have been given short shrift in bioethics.

So far I have surveyed the history and development of the idea of consensus and its analytic dimensions. A number of themes have emerged that will return in various guises when we address consensus in relation to ethics committees. In order to achieve a transition to that more narrowly targeted inquiry, we must somehow reckon with the move from "macrosociology" to "microsociology". Simply put, the sort of consensus that primarily concerns me in this paper is that which occurs in a special kind of small group, rather than in society at large. Therefore, it will be important to canvass some techniques for the assessment of small group relations before moving on to consensus in ethics committees.

V. SMALL GROUP DECISION-MAKING AND APPLIED SOCIAL SCIENCE

A common criticism of small group deliberation is that it can too easily be infected by social pathologies, leading to a distorted "groupthink" [10]. While often well-founded, one should not infer from this criticism that small groups can never produce high quality assessments of problematic situations. There is an important confusion behind this error, namely, that the interpersonal influences in a small group must create obstacles to thinking clearly together. But from the fact that each small group has a specific interpersonal structure, one cannot infer faulty deliberative processes. Instead, by learning about the structure of a particular group one may be able to design interventions that can improve the quality of the group work. The ethics committee need be no exception.

Concerning the importance of understanding the structure of a small group in order to intervene in its processes, analytical techniques provided by J.L.Moreno [14] provide an illuminating example. Moreno developed the science of "sociometry" to analyze the structures of small groups according to patterns of interpersonal choice and rejection. These structures can be exhibited on a "sociogram", a sort of map of interpersonal relations. For our purposes, sociometry is especially useful as it can exhibit group cohesiveness, which as we have noted above is taken to be causally related to consensus.

According to sociometric theory, an earmark of cohesion in a small group is the presence of linked triads (a group of three in which each chooses the other two), known as "chains". In general, as there are more unlinked triads, more dyads (a couple in which each chooses the other), or more non-chosen individuals (or "isolates"), there is less cohesion. The mere presence of a "star", or frequently chosen group member, is in itself no evidence of cohesion, or even of power. For example, a star might be chosen by every group member but none of the group members chooses each other (a fairly common phenomenon among certain groups, such as mental patients who choose only the therapist), so this would not be considered a highly cohesive group. Or a star may be in a mutual dyad with only one other member, who is otherwise an isolate. Overshadowed by the star, this "Rasputin" can become the true power center of the group.

A sociogram can also be used in the prediction of the result of social processes and, if desired, as a guide for active intervention in the reconstruction of a group. Short of that, a sociometric study can identify potential trouble spots in a group that may be amenable to procedural safeguards. To illustrate the usefulness of sociometric investigation with regard to ethics committees, I have created a sociogram of a hypothetical ethics committee in which each member has been asked to choose or reject up to three other members according to the criterion: "With whom would you like to discuss a difficult case?" (See Figure 1)

Note first the powerful triadic structures linking four of the nine committee members. These sub-systems are even more formidable when one considers that the four most chosen committee members are also members of this chain: A (nine choices), E (five choices), and B and F (four choices). As the universally-chosen chairperson on this criterion, A undoubtedly exerts considerable influence. But notice that E, the second most chosen member, may not be as effective within the group as that rank might suggest, for he actively rejects C, the surgeon. By contrast, although B, the psychiatrist, received fewer choices, the facts that he does not actively reject anyone and that he has access to the other four as well as to D (*via* a mutual dyad), suggest a relatively more important role.

As a potential sub-group the women in this hypothetical ethics committee are highly disorganized. While all choose A none are reciprocated. J, the social worker, cannot even muster three choices and receives only one, as well as one rejection. The administrator, I, selects A, F, and H and is either not reciprocated or is actively rejected. H is in a very weak position by virtue of only being chosen by J and the isolated I. Yet the surgeon, C, is the most isolated member: he receives no choices

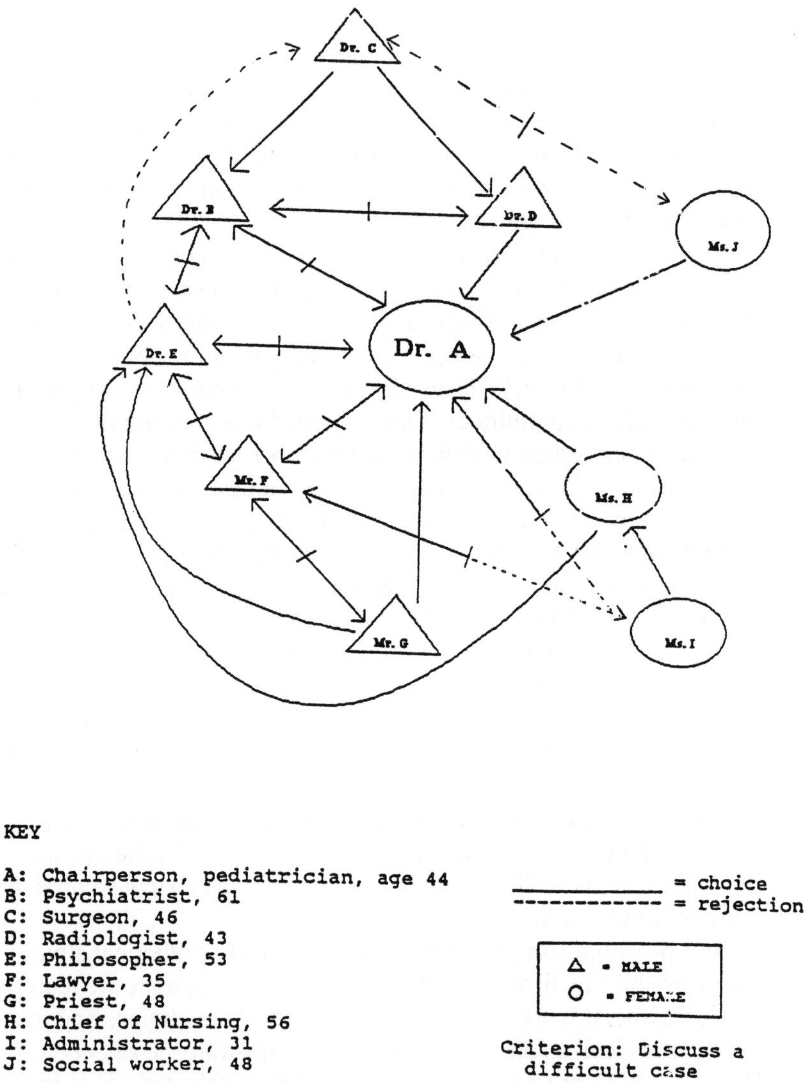

Fig. 1. Sociogram of a hypothetical hospital ethics committee.

and two actively reject him.

The sociogram can be useful in managing the committee's affairs. For example, the popularity of the chairperson should be monitored so that her formulations do not shape the discussion without some critical assessment. The exclusion of a substantial minority from the tightly organized power structure should be addressed, perhaps by encouraging them to articulate their views early on in deliberating about a particular issue. If subcommittees are ever in order, it will be easy to select from the triads for sheer efficiency.

Finally, the sociogram will be most useful if the members of the committee are prepared to overcome their potential embarrassment and resentment at having their relationships so graphically illustrated, for then they will be aware of the interpersonal realities that could inappropriately affect their group processes. Work in applied social science has also contributed specific procedural protocols to improve the results of small group decision-making. For example, there is social psychological evidence that the quality of consensus decision-making can be enhanced if the process is rational and systematic [6]. This is said to be necessary because groups are capable of unanimously agreeing on an incorrect solution to a problem, mainly because they have failed to consider all feasible alternatives.

What is known as a "vigilant" group decision-making strategy includes the following steps: obtain as much information as possible about the alternatives; thoroughly discuss the value of each alternative; after evaluating the alternatives rank them according to the most and least desirable; if there is an unranked middle-range of alternatives, discuss and rank them; and systematically reconsider the rank assigned to each alternative, without hesitating to change a rank if that is warranted [6]. Clearly this sort of systematic procedure would have to be adapted to the unique role of the ethics committee. For instance, one alternative the committee will want to consider and rank from time-to-time is that of giving no advice at all.

It is perhaps worth noting that similar procedures for "quality control" of small group decision-making can be identified on epistemological grounds. Distinguishing between group consensus as *constitutive* of the truth of a proposition or rather as *evidence* for that proposition, I have argued [11] that before adjourning a group would be well-advised to consider of its apparent conclusion: Have all relevant points of view been taken into account? Have all rational considerations been taken into account? Put in this way, no matter which kind of epistemologist of consensus one is, constitutive or evidentiary, one can be satisfied that the essential condition for a satisfactory conclusion has at least been recognized.

VI. PROBLEMS OF CONSENSUS DECISION-MAKING IN THE ETHICS COMMITTEE

One can distinguish three different sorts of institutional committees in the American health care system that have more or less explicitly ethical charges concerning human beings: Institutional Review Boards (IRBs) that monitor the use of human subjects research; Infant Care Review Committees (ICRCs) that consider questions about the care of newborns, especially in neonatal intensive care units; and Hospital Ethics Committees (HECs), which are mainly concerned with ethical questions in adults and older children. These distinctions are important because it is my hypothesis that consensus has more of a role in these panels as one moves from IRBs to ICRCs and then to HECs. IRBs are responsible for ensuring that the statutory requirements of specific Federal legislation have been respected. In this legal capacity clear-cut majority views become more important than in ICRCs, which are responsive mainly to rather general federal requirements of states to investigate complaints of non-treatment of a handicapped infant. Since HECs are not currently associated with government regulation of health care, except in the State of Maryland which requires them of all hospitals, they have still more flexibility than ICRCs to determine their own functions, including the manner in which their collective views are to be identified. In a formal sense, the vast majority of institutional ethics committees regard their role as "advisory"; rarely are their findings mandatory for the parties to a controversy.

Regardless of the status of an ethics committee's recommendations, whether advisory or mandatory, functionally the committee must arrive at some conclusion. Even if a committee demurs from giving advice in a certain case, that is still a decision that the group has reached. In an informal sense, it is evident that under certain political conditions the supposedly advisory nature of an ethics committee's views could be difficult for the principals to ignore.

Evidence from survey data indicates that most ethics committee members characterize their approach in terms of developing a consensus. In the most comprehensive comparative study of ethics committees in three American states [7], from 75 percent to 92 percent of committee members reported that decisions are made by means of consensus, with most of the remainder reporting that votes are taken. Of course, what precisely is meant by consensus in this context is not entirely clear. There are at least three possible meanings: (1) the positive view of all or virtually all individual committee members; (2) the product of an effort to accommodate mutually the views of all or virtually all committee members (a sort of coordination problem); or (3) an attempt by the committee members to replicate what all or virtually all of them think would be morally acceptable to the relevant larger community. This ambiguity raises interesting empirical questions about the self-perception of ethics committee members as they engage in their activities.

I argue that the crucial philosophical problem in the concept of an ethics committee is the ambiguous moral status of consensus ([11]; [12]). According to the predominant contemporary assessment of the relation between fact and value, from the fact that a number of individuals have found one proposition or another to be morally sound, the moral soundness of that proposition cannot be inferred. Related to this logical limitation, rooted deep in Western philosophy is the suspicion that, empirically, moral truth is in fact less likely to be achieved by groups, which are vulnerable to the corruptions of political processes and interpersonal dynamics, than by well-informed and reflective individuals. I call this standard philosophical critique of the moral authority of consensus, with its logical and empirical elements, the epistemological view.

Even if we grant the epistemological view its point in a formal sense, it is hard to see how human institutions could proceed without routinely relying on consensus. This routine reliance on consensus can be said to persist on a first-order level and a second-order level. The machinery of organized social life, in all its dealings, surely requires the lubricant of common agreement. Further, in a second-order sense, there is also a social need to see that collective opinion as by-and-large right opinion. The analysis of the legitimation processes of consensus views is normally the province of sociology, which concerns itself, for example, with the formal and informal processes through which bodies such as ethics committees gain or lose their perceived legitimacy in the institution. Thus I call the position that emphasizes the social and political realities of consensus decision making, and does not trouble itself with the traditional philosophical critique of consensus, the sociological view.

These philosophical and sociological assessments of consensus decision making suggest that social contract theory might offer a way to ground the ethics committee in rationality. Placed behind a Rawlsian "veil of ignorance", hypothetical ethics committee members may attain a disinterested moral point of view. Unfortunately, as I have noted in pursuing this argument in detail elsewhere [13], multiple principles of justice can be identified that give different results from behind the veil. Furthermore, as Veatch [21] points out, there are a number of practical limits to this attempt at social contract analysis. For example, members of ethics committees who are health care providers come from both a Hippocratic heritage and a clinical perspective. There are also considerable variations from one clinical consensus to another, with different ethics committees evaluating the same case differently. Finally, ethics committees in health care institutions affiliated with particular religious groups ought in fact to reflect the "value commitment" of the sponsor. Contract theory provides no basis for concluding that the ethics committee's deliberations take precedence over the reflections of the competent patient concerning his or her own care. By contrast, even if one takes into account the limitations of the analogy between veiled decision makers and ethics committee members, social contract theory arguably provides a basis for holding the surrogate to a "limit of reasonableness" [20] in making decisions for a patient. But these limits will rarely be at issue when an informed surrogate has been duly selected by the patient.

VII. DYNAMIC VERSUS STATIC CONSENSUS

The foregoing discussion of consensus in bioethics has mainly considered it in relation to decision-making strategies in small groups, like institutional ethics committees, which might be called "dynamic consensus". But the last remarks about bioethics as a form of practical wisdom or social intelligence move toward a conception of consensus as a common point of view held more or less independently by large numbers of well-informed citizens, which might be called "static consensus".

To be interested in static consensus is to believe that one can inquire in an abstract but meaningful way about the consensus among, say, physicians concerning such-and-such a question, regardless of their institutional sitting or societal origin. The implication is that this consensus among professionals, if there is one, is interesting not because they have all been able to sit together and work it out, but as a more-or-less typical attitude that has emerged from the experience of this group in the practice of the discipline. These attitudes can be discovered in a survey that elicits responses from a targeted population. Though it is not specifically concerned with values, one should mention in this context Charles S. Peirce's notion of the "long run" of scientific inquiry, according to which the truth is defined as that opinion which the community of inquirers is fated to reach if science proceeds long enough [16].

Static consensus on value questions among professionals does not raise the interesting questions of interpersonal processes discussed earlier in this paper. But this form of consensus deserves mention if only because the collective opinions of experts are often highly influential in the creation of public policy. In the area of human reproduction, the question of the moral significance of static professional consensus was dramatically raised in a cross-cultural study of ethical attitudes among geneticists [5]. Among the results from 643 respondents, Fletcher and Wertz found "strong consensus" on certain propositions, "moderate consensus" on others, and "no consensus" in other areas. For example, there was strong consensus that the protection of the mother's confidentiality overrides disclosure of true paternity, moderate consensus that genetic screening in the workplace should be voluntary, and no consensus on the confidentiality of a diagnosis of Huntington's disease.

Consider by contrast the dynamic processes of consensus formation, flawed as they may be, that characterize small deliberative groups like committees and commissions. Imagine that a hospital ethics committee is asked by a physician for guidance on the question of disclosure of a diagnosis of Huntington disease to the pregnant daughter of a newly deceased patient. The deceased patient's wife opposes disclosure, but the physician notes that the pregnancy is still in its early stages. Although the outcome of the committee's deliberation can hardly be predicted in advance, the recommendation to inform, not to inform, or not to take a position (which amounts of course to the same thing as noninforming), has profound implications for actual individuals. With so much at stake, any such body with even a minimal sense of responsibility will surely engage in a lively discussion of the issue. I would argue that interactive deliberation, impelled partly by the fact that real and significant consequences could follow, gives dynamic consensus a greater claim to moral significance than the static consensus that can be gleaned from a survey of discrete individuals.

Let me be clear that I do not think there is any moral superiority *a priori* to conclusions reached dynamically, in a small group like a committee or commission. Rather, my argument is a pragmatic one, to the effect that conversation is far more likely to produce a reasonably thorough examination of the morally relevant considerations than the independent responses of separated individuals. All other things being equal, I would put my money on the conclusion reached through a candid and open discussion among moral equals.

Finally, early in this paper I asserted that the theoretical nature of moral consensus is by-and-large identical in the case of ethics committees and commissions. It is now appropriate to note one obvious exception to this generalization. Commissions established by organizations governed on a democratic basis, whether governments or professional groups, arguably carry an added burden that is a familiar conundrum of representative democracy: should the commissioners represent their own views only or should their consensus reflect that which they perceive to be the generic view of their constituents? Since commissioners are not legislators, their burden in this respect seems to be lessened, and their corresponding obligation to be guided more by their own dynamic group process rather than by notions reached through static consensus appears to be greater. Nevertheless, occasions will surely arise when the integrity of a select commission's conclusion is out-of-step with that of most members of the polity or of the authorities responsible for its appointment. This is all to the good, for if this were this not the case the commission process would be a charade, merely a rubber-stamp for a popular, static consensus.

VIII. CONCLUSION

So long as one concentrates only on the similarity of static and dynamic consensus, namely, that they both eventuate in what Plato called "mere opinion", one misses the salient fact that the process according to which consensus is reached conditions the authority of the conclusion. Thus, as I have noted, although the political and social pitfalls associated with small-group decision-making are evident, a number of techniques for intervention are available to improve the process. I contend that, in principle, dynamic consensus has a legitimate claim to a degree of moral authority that cannot be attributed to static consensus.

I am nearly tempted to conclude that these two sorts of consensus, one that emerges from group interaction and one that is merely a coincidence of opinion, are so different that they should not even be brought under the same general term. Perhaps the greatest problem with static consensus is that it makes it easy to continue to use the word "consensus" as though the goal of inquiry is simply agreement. By contrast, in the dynamic context of give-and-take among equals, consensus is a condition of the evolving process, one that climaxes in a morally satisfactory result. When this is the case moral inquiry is an integral part of the life of an institution, as it should be. Thus we have returned to the organic conception of social life that was home for the earliest recorded reflections on consensus.

Division of Humanities in Medicine
State University of New York Health Science Center at Brooklyn
Brooklyn, New York, USA

NOTES

[1] The author expresses his gratitude to his fellow participants in the conference on "Technische Eingriffe in die menschliche Reproduktion: Perspektiven eines moralischen Konsenses", who provided numerous helpful criticisms of a previous draft.

BIBLIOGRAPHY

1. Arras, J.: 1991, "Getting Down to Cases: The Revival of Casuistry in Bioethics," *The Journal of Medicine and Philosophy* 16, 29–52.
2. Beauchamp, T. and Childress, J.: 1989, *Principles of Biomedical Ethics*, 3rd ed., Oxford University Press, New York.
3. Clouser, K.D. and Gert, B.: 1990, 'A Critique of Principlism', *The Journal of Medicine and Philosophy* 15, 219–236.
4. Dewey, J.: 1958, *Experience and Nature*, Dover Publications, New York.
5. Fletcher, J.C. and Wertz, D.C.: 1987, 'Ethics and Human Genetics: A Cross-Cultural Perspective', *Seminars in Perinatology* 11, 224–228.

6. Hirokawa, R.Y.: 1984, 'Does Consensus Really Result in Higher Quality Group Decisions?', in G.M. Phillips and J.T. Wood (eds.), *Emergent Issues in Human Decision Making*, Southern Illinois University Press, Carbondale and Edwardsville, Illinois, pp. 40–49.
7. Hoffman, D.: 1990, Study of Hospital Ethics Committees in Maryland, the District of Columbia, and Virginia; Preliminary Results, University of Maryland School of Law, Baltimore, Maryland, personal communication with Diane Hoffman.
8. Jonsen, A.R. and Toulmin, S.: 1988, *The Abuse of Casuistry*, University of California Press, Berkeley, California.
9. Key, V.O.: 1961, *Public Opinion and American Democracy*, A.A. Knopf Publishers, New York.
10. Lo, B.: 1987, 'Behind Closed Doors: Problems and Pitfalls of a Hospital Ethics Committee', *New England Journal of Medicine* 317, 46–49.
11. Moreno, J.D.: 1988, 'Ethics By Committee: The Moral Authority of Consensus', *The Journal of Medicine and Philosophy* 13, 411–432.
12. Moreno, J.D.: 1990, 'What Means this Consensus? Ethics Committees and Philosophic Tradition', *Journal of Clinical Ethics* 1, 38–43.
13. Moreno, J.D.: 1991, 'Consensus, Contracts, and Committees', *The Journal of Medicine and Philosophy* 16, 393–408.
14. Moreno, J.L.: 1953, *Who Shall Survive?*, Beacon House Publishers, Beacon, New York.
15. Partridge, P.H.: 1971, *Consent and Consensus*, Praeger Publishers, New York.
16. Peirce, C.S.: 1955, 'How to Make Our Ideas Clear', in J. Buchler (ed.), *Philosophical Writings of Peirce*, Dover Publications, New York.
17. Rawls, J.: 1980, 'Kantian Constructivism in Moral Theory: The Dewey Lectures 1980', *Journal of Philosophy* 77, 515–572.
18. Stanley, M.: 1978, *The Technological Conscience*, Free Press, New York.
19. Toulmin, S.: 1986, 'How Medicine Saved the Life of Ethics', in J.P. DeMarco and R.M. Fox (eds.), *New Directions in Ethics*, Routledge and Kegan Paul Publishers, New York, pp. 265–281.
20. Veatch, R.M.: 1984, 'Limits of Guardian Treatment Refusal: A Reasonableness Standard', *American Journal of Law and Medicine* 9, 427–468.
21. Veatch, R.M.: 1990, 'Ethical Consensus Formation in Clinical Cases', (unpublished manuscript).
22. Walters, L.: 1989, 'Commissions and Bioethics', *The Journal of Medicine and Philosophy* 14, 363–368.

JAMES F. CHILDRESS

CONSENSUS IN ETHICS AND PUBLIC POLICY: THE DELIBERATIONS OF THE U.S. HUMAN FETAL TISSUE TRANSPLANTATION RESEARCH PANEL

I. INTRODUCTION

Should the U.S. federal government provide funds for scientists to pursue research on the treatment of various diseases, such as Parkinson's disease, using tissue from deliberately aborted fetuses? In November 1989, Dr. Louis Sullivan, the Secretary of the Department of Health and Human Services (DHHS), announced the indefinite extension of the moratorium (in effect since March 1988) on the use of federal funds in such research, despite the majority recommendation (usually described as a "consensus") of the Human Fetal Tissue Transplantation Research Panel and the unanimous recommendation of the Advisory Committee to the Director of the National Institutes of Health. In this essay I want to explore the role of consensus in the deliberations of the Panel, on which I served as a member, and in the broader societal debate and policymaking – both to illuminate the ethical and political controversy in the U.S. about human fetal tissue transplantation research (hereafter HFTTR) and to explore some issues surrounding consensus on national governmentally-appointed bodies to examine ethical and policy questions in biomedicine.

This indefinite moratorium was part of a long pattern of federal governmental avoidance of research connected to the fetus, including research on reproductive technologies ([31], pp. 130–147, Testimony of John Fletcher). However, there are good reasons for sharply distinguishing the controversy about transplantation research using the tissue of dead fetuses following induced abortion from research on living embryos or fetuses in utero or ex utero. And it is possible to support the

Kurt Bayertz (ed.), The Concept of Moral Consensus, 163–187.

use of cadaveric fetal tissue regardless of one's views about the status of the fetus and the morality of abortion.

It is not surprising that, when the question arose about the use of federal funds for HFTTR, following deliberate abortions, DHHS sought independent advice through advisory committees or panels to include participants from the public and diverse professional backgrounds. "Since the early 1970s [in the U.S.]," notes Patricia King ([18], p. 249), "complex ethical, social, legal, and scientific controversies generated by scientific and medical advances have been referred increasingly to national commissions, committees, boards, or panels." According to King, who has participated in several of these groups, including the HFTTR Panel, they offer several advantages over courts, legislatures, and regulatory agencies, including greater flexibility and more extensive analysis in approaching complex social dilemmas, the possibility of explicit and self-conscious incorporation of "ethical premises into their deliberations," and a successful track record of reaching consensus in their advice and recommendations in many instances ([18], p. 250). The following case study provides key facts about the HFTTR Panel for purposes of exploring and evaluating its efforts to reach consensus in its deliberations, in part as a way to identify larger ethical and policy issues about consensus.

II. CASE STUDY[1]

Because of promising results in animal research in the U.S. and other countries, researchers at NIH were interested in experimentally transplanting human fetal neural tissue, following elective abortions, into patients with Parkinson's disease. In late 1987, the Director of NIH, who had the legal authority to approve the protocol for HFTTR, sought further review from the office of the Secretary of DHHS, in part because the proposed research had "the potential for publicity and controversy" [36]. (NIH had long provided funds for research involving the use of fetal tissue, e.g., to develop vaccines, and had recently awarded an extramural grant for research involving the transplantation of human fetal pancreatic cells into patients with diabetes.) In March, 1988, the Assistant Secretary for Health, Robert Windom [35] declared a moratorium on the use of federal funds in HFTTR until "special outside advisory committees" could be formed to hear testimony, deliberate, and offer their recommendations to NIH, in response to ten questions, which he describes as "primarily ethical and legal" and which focus largely on the connection or linkage between HFTTR and abortion.[2] Then in early summer, 1988, NIH appointed the Human Fetal Tissue Transplantation Research Panel to meet in the fall of 1988 to respond to the Assistant Secretary's ten questions and then to submit its finished report to the NIH Director's Advisory Committee, which then would advise the Director and through him DHHS. Judge Arlin Adams, a retired federal judge from Philadelphia and a Republican who is generally opposed to abortion, was appointed to head the Panel.

The HFTTR Panel convened in a public meeting on September 14–16, 1988, to hear testimony on scientific, legal, and ethical views from over fifty invited speakers and over fifteen representatives of public interest groups. Although the Panel had originally been expected to complete its deliberations in the scheduled three days of meetings, that time was inadequate, and a second meeting was arranged for October 20–21 to consider a draft report that offered relatively brief "responses" to the Assistant Secretary's ten questions without providing substantial justification for the responses. After some discussion about the possibility of circulating and approving justificatory statements without a third meeting, the Panel decided to submit only what had been developed and accepted by the end of the second meeting. However, a few minutes before the second meeting adjourned, two Panel members, James Bopp and James Burtchaell, distributed to the Panel a substantial dissent to the report. Panelists in the majority later expressed their concern that such a long and eloquent dissent would simply smother the report's brief responses, which lacked adequate justificatory statements, and they requested a third meeting, which was scheduled for December 5, when the Panel could consider drafts, prepared by different panelists, of "considerations" for each "response" to the ten questions and could put the report into final form. In addition to the "responses" and "considerations," the final report contains three concurring statements, including a long one signed in whole or in part by eleven panelists, and dissenting statements by four panelists. A second volume of the report contains the written testimony received by the panel.

The Panel concluded that it is "acceptable public policy" for the federal government to support HFTTR, using tissue from deliberately aborted fetuses, as long as some "appropriate guidelines" or "safeguards" are in place, particularly to separate as much as possible the pregnant woman's abortion decision from her decision to donate fetal tissue. These "guidelines" or "safeguards" include prohibiting financial remuneration to women who provide fetal tissue (in order to avoid a financial incentive for abortion); prohibiting the woman's designation of a patient as a recipient of the fetal tissue (in order to avoid a motivation of specific altruism for a loved one to have an abortion to provide fetal tissue); not offering preliminary information about tissue donation until after the abortion decision (unless specifically requested); not promising that the fetal tissue could or would be used; not modifying the timing or method of abortion because of the potential use of the fetal materials; informing potential recipients of such tissues and research and health care participants of their source; and according the "same respect" to human fetal tissue accorded to "other cadaveric human tissues entitled to respect" ([23], pp. 1–22).

The HFTTR Panel's report was submitted on December 14, 1988, to the Advisory Committee to the Director of NIH and representatives of the National Advisory Councils to NIH, with oral presentations by nine of the ten Panel members who attended (another absent panelist's statement was entered into the record). The Advisory Committee ([2], p. 4) quickly concluded that the Panel's report was "clearly an impressive and skillfully crafted document," which reflected "extensive and thoughtful work," and that "given the divisiveness underlying our society on the issues related to the topic under consideration, the report represented *a remarkable consensus*" (emphasis added). The Advisory Committee ([2], p. 4) "further concluded that the consensus of the Panel reflected the consensus of the country itself, where widely divergent views are held about the morality of elective abortions and about the use of fetal materials derived from such abortions for the purposes of research." After reviewing and discussing the Panel report, the Advisory Committee unanimously accepted the report and recommendations of the HFTTR Panel as written, recommended that the Assistant Secretary for Health lift the moratorium on the use of federal funds in HFTTR that uses tissue from induced abortions, and recommended, within the current laws and regulations governing research using human fetal tissue, the development by NIH staff of "additional policy guidance" to implement the HFTTR Panel's guidelines ([2], p. 7).

III. CONSENSUS AND THE HFTTR PANEL'S DELIBERATIONS

Consensus: Outcome or Process? Consensus (from the Latin consentire – to feel with) can be viewed in terms of an outcome or process. As Peter Caws ([7], p. 377) notes, even though "consent" and "consensus" share the same etymological root and much the same content, the former focuses on the act of agreement, while the latter focuses on the fact or substance of agreement. Thus, "consensus" tends to refer to outcome, the group's agreement, and that is how the Advisory Committee viewed the consensus reflected in the report offered by the Panel.

A science writer who observed the HFTTR Panel's meetings describes the Panel's process as one of seeking consensus:

> Despite the diversity of views held by members of the ad hoc panel, the group steadfastly tried to follow a consensual approach during its deliberations. Although consensus was difficult to achieve, the panel members consistently tried to accommodate one another's respective positions. Thus, in most cases, very disparate philosophical positions were melded into a coherent stance that was deemed acceptable by a substantial majority of the panel. However, neither of these observations should be taken to suggest that the debate within the panel was somehow constrained by the majority viewpoint, as indeed it was not ([14], A1).

Much of the Panel's discussion was devoted to debating and modifying the formulation of particular responses, in an effort to gain as much consensus as possible. Often, however, a small minority of one to three voted against the carefully-constructed compromises.

In short, the panelists sought as much consensus as possible on the formulation of various responses to the Assistant Secretary's questions, but then they voted on each question, listing the totals for, against, and abstaining. Even though claims about the Panel's consensus became important in subsequent policy debates, the numbers actually voting for and against particular responses also became important, as was evident in the hearings about HFTTR in April 1990 ([31], pp. 66, 81). And the dissents published along with the majority report and concurring statements played an important role in the reception and subsequent use of the report.

Closure of a debate by consensus can be distinguished from other types of closure. H. Tristram Engelhardt and Arthur Caplan ([13], pp. 13–16) note that scientific controversies with heavy moral and/or political overlays may be brought to closure through (1) loss of interest, (2) force, (3) consensus, (4) sound argument (with several subsets), and (5) negotiation. The first three modes of closure are "achieved neither by reliance upon sound arguments nor by reliance upon fair principles of negotiation" ([13], p. 16). Although an agreement that results from force or consensus may be correct, it is not based on reason. Consensus need not explicitly rely on the procedural fairness required in negotiations, but "the termination of a controversy by consensus is 'fair' in the sense of respecting the views of those involved" ([13], pp. 16–17). A "community of belief" results, whatever its influences.

The conclusions reached by the HFTTR Panel roughly represent a consensus in contrast to the other ways of bringing closure. However, the Panel did use a procedure of voting, which some commentators distinguish from less formal approaches that produce consensus (e.g., the Quaker "sense of the meeting"). Furthermore, all the other modes of closure, with the exception of force, were evident in the Panel's deliberations – loss of interest in pursuing points, sound argument, and, of course, negotiation.

Extent of Agreement Necessary for a Consensus. Within the boundaries of a group, there is a question about how many of its members have to agree before their agreement is appropriately described as a "consensus." Clearly, if there is unanimity, there is consensus, but, in the absence of unanimity, how many must affirm a position before it represents a consensus? As noted above, the Advisory Committee ([2], p. 4) indicated that the Panel's "consensus" reflected "the consensus of the country." However, it is unclear how there can be a "consensus of the country" that HFTTR is "acceptable public policy" with certain "safeguards," which is what the Panel held, if there are in the country "widely divergent views ... about the use of fetal materials derived from such abortions for the purposes of research" ([2], p. 4). What the Advisory Committee means by "consensus" is not clear.

Even though a strong majority of the HFTTR Panel supported the responses to each question, as many as three out of twenty-one dissented from the responses to the major ethical questions. Furthermore, four panelists filed dissents to the report as a whole. While it is possible to say that the report reflected the "general consensus" of the Panel, the language of "general consensus" may refer to the breadth or extent of the agreement or to the nature of the agreement, including its level of generality. Ambiguities surround the language of "substantial consensus" ([2], p. C 5, Statement by LeRoy Walters).

The Nature of the Consensus Reached: What or Why? A consensus may be limited to the conclusions or may extend to the reasons for the conclusions. For instance, there could be a consensus about what policies should be adopted without a consensus about why those policies ought to be adopted. This discrepancy occurred on the HFTTR Panel; panelists in the majority accepted the "guidelines" or "safeguards" to separate the abortion decision as much as possible from the donation decision for several different reasons. Some panelists viewed these safeguards as morally required because abortion is usually immoral. Other panelists accepted the safeguards against commercialization and designation of recipients of the donated fetal tissue and the like (1) in order to avoid fanning the flames of the abortion controversy in a seriously divided society, and/or (2) in order to reduce the likelihood of harm to or coercion and exploitation of women who might be pressured into having an abortion in order to provide fetal tissue.

In exploring the nature of the consensus reached, it is necessary to note the different kinds of questions raised by the Assistant Secretary for Health for consideration by the Panel. Some of the questions requested *empirical information* (e.g., what actual steps are involved in procurement of fetal tissue); some asked for an *interpretation of the law* (e.g., whether state laws regarding the use of cadaveric fetal tissue would apply to transplantation research and whether obtaining informed consent for the donation of fetal tissue from a pregnant woman planning an abortion would constitute a prohibited "inducement" to abortion). Some asked for *medical and scientific predictions* (e.g., the likelihood that transplantation using fetal cell cultures will be successful and will obviate the need for fresh fetal tissue within a certain time frame), while others asked for *social predictions* (e.g., whether HFTTR would encourage women to have abortions they would not otherwise have had, and what impact the common use of aborted fetal tissue might have on activities and procedures of abortion clinics). Still others asked for *moral or ethical evaluation*, such as determining the "moral relevance" of the fact that the tissue for transplantation comes from deliberately aborted fetuses. Several questions cut across these categories; for example, the question whether maternal consent is a sufficient condition for the use of fetal tissue could be seen as legal and/or moral. Medical and scientific facts have to be combined with standards of moral evaluation to answer the question about whether enough animal studies have been performed to justify proceeding to HFTTR for some diseases. And the question about the possible prohibition of the donation of fetal tissue between known persons involves judgments about morality and feasibility as well as whether such a prohibition would jeopardize the likelihood of clinical success. Even though sharp distinctions between factual and evaluative statements are difficult to defend, it is important to note the different sorts of claims being made in the responses to the questions and thus the different kinds of consensus involved. In what follows I will concentrate on the moral and ethical questions and on the mixed questions.

The Moral Relevance of Induced Abortion to HFTTR. Following is the first and perhaps the most fundamental question raised by the Assistant Secretary: "Is an induced abortion of moral relevance to the decision to use human fetal tissue for research? Would the answer to this question provide any insight on whether and how this research should proceed?" After much debate, the Panel answered (18 yes, 3 no, 0 abstain) that "it is of moral relevance that human fetal tissue for research has been obtained from induced abortions," but that it is also "*acceptable public policy*" to use such tissue because of the possibility of relieving suffering and saving life ([23], p. 1). However, the fact of "induced abortion creates a set of morally relevant considerations," and the Panel derived its guidelines from the society's deep moral convictions about abortion. These guidelines attempt to separate the decision to abort and the procedures of abortion as much as possible from the retrieval and use of fetal tissue. I will return to the Panel's answer to this question later because the decision not to say "*ethically* acceptable public policy" may have had fateful consequences; at the very least, it provided ammunition for the critics of HFTTR.

The Significance of Framing Problems. How problems are framed is very important for the deliberations of any group. The Assistant Secretary's ten questions stressed the linkage or connection of HFTTR with abortion decisions and practices. This linkage is understandable, because without HFTTR's connection to abortion the Director of NIH would have approved the research without involving the Assistant Secretary for Health. However, in focusing so exclusively on abortion-related issues, the ten questions severely constrained the Panel's deliberations. An alternative set of questions could have been more neutral or could have started from the well-established societal practice of using cadaveric human tissue in research, education, and transplantation – cadaveric fetal tissue has already been used in all these ways too (though not widely for transplantation). Then the Panel could have extended the principles embedded in these practices and moved by analogical reasoning to the use of cadaveric fetal tissue in transplantation research, probing for consistency between this proposed use in transplantation research and the established principles and paradigm cases. While it is very doubtful that an alternative framing of the problem would have produced greater consensus, the final report would have been very different and perhaps more persuasive. The linkage to abortion would have been addressed secondarily in trying to determine whether the cause of death makes a difference in the use of the cadaveric tissue.

Implications of Views about the Status of the Fetus and the Morality of Abortion. Views about the status of the fetus and the morality of abortion clearly were significant in the judgments of individual panelists. The fetus can be viewed as mere tissue, as potential human life, or as full human life. A proponent of the first interpretation of fetal life would probably view the donation of fetal tissue as analogous to living donation of an organ, such as a kidney, or tissue, and would not be inclined to respect fetal tissue any more than a person's excised appendix. The other two views of the status of the fetus do not appear to entail any particular position on the use of fetal tissue in transplantation research, because it is considered legitimate to use the tissue of adult cadavers who have been recognized as full persons. The Panel's recommendation of equal respect for fetal tissue could be based on either the full or the potential humanity of the fetus or on avoiding offense to others. Simi larly, it is possible to oppose abortions and still accept HFTTR without any inconsistency, and also to accept the safeguards to separate abortion decisions from donation decisions without implying that abortion is generally immoral. Panelists in the majority appeared to hold one of two positions: (1) abortion is morally acceptable and HFTTR using aborted fetal tissue is morally acceptable, or (2) abortion is "immoral and undesirable," even though legal, but HFTTR can be morally separated from abortion decisions and practices that in fact produce the fetal tissue ([23], p. 2). Perhaps the Panel could have usefully explored the possibility of common ground on some exceptions to the moral prohibition on deliberate abortion affirmed by some panelists. For instance, there is widespread agreement about the moral justifiability of abortion in cases of ectopic or fallopian tube pregnancies, but it is not clear that much usable tissue can be obtained following such abortions.

Debates about Altruistic Incentives for Abortion. The Panel's moral debate focused to a great extent on predictions about whether government-supported HFTTR would lead some pregnant women to have abortions they would not otherwise have had. These predictions depend in part on interpretations of why and how pregnant women choose to have an abortion, which should be resolvable by empirical data. But rather than conducting a careful study (for which there was no time), or even systematically examining the relevant literature, panelists mainly reported their impressions, sometimes sexist in tone ([18], p. 253). The Panel concluded "that the reasons for terminating a pregnancy are complex, varied, and deeply personal" and regarded "it highly unlikely that a woman would be encouraged to make this decision [to abort] because of the knowledge that the fetal remains might be used in research" ([23], p. 3). Furthermore, the majority held that the recommended "guidelines" or "safeguards" would reduce the likelihood of an impact of HFTTR on the incidence of abortion. In recommending the maximum possible separation of the decision to abort and the decision to donate, the Panel noted that its members "take this stand either because they do not want to do anything that might encourage abortion or as a concession to those who do not want to risk encouraging abortion" ([23], p. 4). Here again there was consensus on what without consensus on why.

Opponents of HFTTR stressed that women's decisions to abort are often ambivalent and sometimes altruistic, and that the possibility of donating fetal tissue to benefit another human being could reasonably be expected to "tip the balance in favor of abortion for some women who are ambivalent" ([5], pp. 56–57). Even though the Panel's recommendation of anonymity between donor and recipient would eliminate the possibility of specific altruistic donations for a family member, the Panel could not rule out the possibility that general altruistic motives would lead some ambivalent women to choose an abortion they would not otherwise have chosen: "knowledge of the possibility for using fetal tissue in research and transplantation might constitute motivation, reason, or incentive for a pregnant woman to have an abortion" ([23], p. 4). However, the majority concluded that it was justifiable for the society through the federal government to take this risk. This point was stated most clearly in a concurring statement by John Robertson (and joined in whole or in part by ten other panelists). Society accepts risks to human life frequently in the pursuit of activities that serve "worthy goals and when reasonable steps to minimize the loss have been taken" ([24], p. 34). It is not defensible to accept a more stringent policy for fetal lives that might be lost in legal abortions than for other human lives. These risks are at best speculative and can be reduced through the proposed guidelines. And, as the concurring statement ([24], p. 35) noted, the society does not encourage deaths from homicide, suicide, or accidents in order to gain organs for transplantation.

James Mason ([19], pp. 17–18), at the time Assistant Secretary for Health, later charged that the concurring statement simply threw out ethical considerations altogether in its argument. In fact, it offered a different balance of ethical considerations than Mason's own risk-benefit analysis, which held that even one additional fetal death would be too high a price to pay for the potential benefits of HFTTR ([19]; [31], p. 80).

Impact of Previous Fetal Tissue Research on Abortion Decisions. The HFTTR Panel held (19 yes, 1 no, 1 abstention) that there is "no evidence" that the use of fetal tissue in research has "had a material effect on the reasons for seeking an abortion in the past" ([23], p. 3) For example, in FY 1987 NIH awarded 116 grants and contracts (estimated at $11,200,000) for research that involved the use of human fetal tissue. In general this research is not therapeutic, and does not involve transplantation. The use of fetal tissue in the development of the polio vaccine is a well-known example of earlier research using fetal tissue. However, some panelists worried that HFTTR would have more impact on abortion decisions in part because it is "more publicized and promising research" ([23], p. 3). And Assistant Secretary Mason contends that a major difference is that the benefits to the recipient are more direct in HFTTR ([31], p. 80). Some commentators [22] make this point by distinguishing using cadaveric fetal tissue to develop a treatment from using it as a treatment. The Assistant Secretary also noted that many more fetuses – perhaps as many as four for each Parkinson's patient – would be required in HFTTR than in other research, which often involves the use of a stable cell line from a few fetuses (as in the development of the polio vaccine) ([31], p. 79).

Donors of Fetal Tissue. Apart from the question of altruistic incentives for pregnant women to have abortions, a major normative dispute on the Panel concerned the locus of the authority to transfer aborted fetal tissue for transplantation research. The Assistant Secretary asked whether maternal consent is a sufficient condition for the use of fetal tissue in transplantation research, and the affirmative answer received lowest number of votes of the answers to any question (17 yes; 3 no; and 1 abstention). The majority viewed maternal consent as both necessary and sufficient, unless the father objects (except in cases of incest or rape). Among the several possible modes of transfer of fetal tissue – donation, abandonment, expropriation, and sales – the Panel affirmed express donation by the pregnant woman after her abortion decision, in part because it is already the primary mode of transfer of cadaveric organs and tissues in the U.S. and thus is "the most congruent with our society's traditions, laws, policies, and practices" ([23], p. 6). The Panel held that a woman's choice of a legal abortion does not legally disqualify her and should not disqualify her from serving "as the primary decisionmaker about the disposition of fetal remains, including the donation of fetal tissue for research" ([23], p. 6). Disputes about the morality of the pregnant woman's decision to abort "should not deprive the woman of the legal authority to dispose of fetal remains. She still has a special connection with the fetus and she has a legitimate interest in its disposition and use" ([23], p. 6). In addition, the dead fetus has no interests that the decision to donate would violate.

By contrast, some opponents of HFTTR contend that the pregnant woman who chooses to have an abortion loses any authority over the fetal remains; her decision to abort means that she has abdicated the role of guardian ([5], p. 47; [31], pp. 18–19). In response, the majority denies that guardianship is an appropriate model for the donation of cadaveric tissue in part because of the absence of cadaveric interests ([24], p. 36). Ironically, the dissent's position could more easily support HFTTR under a model of expropriation or abandonment or presumed donation, because express donation (or sales) would provide additional incentives for abortions.

The Basis and Generality of Consensus on the HFTTR Panel and the National Commission. The question of the generality of consensus – whether principles or casuistical judgments – has been extensively debated in the context of the National Commission for the Protection of

Human Subjects of Biomedical and Behavioral Research in 1974–79, the first major U.S. commission in bioethics, and there are some important points of comparison and contrast between the work of that commission, which focused on several types of research involving human subjects, including fetal research, and the work of the HFTTR Panel, which focused on transplantation research with cadaveric fetal tissue.

Albert Jonsen, an ethicist on the National Commission, and Stephen Toulmin, a philosopher on its staff, have argued that the Commission reached consensus on important issues in research involving human subjects by focusing on specific types of cases rather than by appealing to principles [17]. The commissioners' taxonomic approach included paradigm cases and analogical reasoning. And, in contrast to claims by Alasdair MacIntyre, among others, about contemporary moral Babel, the commissioners' approach identified features "relevant to the moral acceptability or unacceptability of such research" ([27], p. 610). The commissioners agreed on "the relevant moral considerations," even though they did not always achieve unanimity because they weighed these "moral considerations" differently. However, they approached Babel when they moved away from *what* they agreed about to *why* they agreed about it, particularly to principles to justify their judgments. The "locus of certitude," according to Toulmin ([27], p. 612), was in the judgments about particular types of cases, rather than in general principles.

Because he tends to construe moral principles as absolute, invariant, and foundational, Toulmin may overlook other kinds of moral principles actually at work in the Commission's deliberations. For a closer examination appears to disclose general moral considerations, such as fairness, respecting choices, and not harming others, which could appropriately be called principles. Furthermore, some paradigm cases for analogical reasoning are clearly connected with moral principles. For example, the negative paradigm case of the Nazi experiments condemned at the Nuremberg Trials is also connected with the important moral principles articulated in the Nuremberg code. Thus, for the National Commission the principles and the paradigm cases for analogical reasoning were established and connected as a matter of societal consensus.

In contrast to Toulmin's interpretation of the National Commission's consensus on kinds of cases, James Burtchaell holds that the HFTTR Panel had consensus on the "grounds" for reaching a shared judgment but then failed to apply those grounds properly. He contends that the panelists all "consentually (sic.) held" three convictions – the Nuremberg principle about informed consent, the avoidance of governmental complicity through inducement of abortion, and the avoidance of governmental complicity after the fact through institutionalized arrangements with abortionists ([31], pp. 16–20). "On those three grounds we all stood together" ([31], p. 17). However, Burtchaell suggests that the majority misapplied this consensus about grounds to HFTTR because of their prior biases ([31], pp. 16–20, 66–71). His claim about consensus on grounds is controversial, in part because the notion of complicity implies participation in a moral evil, and not all panelists agreed that abortion is generally a moral evil, even if it is tragic, undesirable, etc. For example, as already noted, there were reasons other than the immorality of abortion for adopting the proposed safeguards.

While the HFTTR Panel certainly presupposed wide agreement on some paradigm cases (e.g., Nazi experimentation) and on the principles that help to define those cases (e.g., the Nuremberg code) in research ethics, there was sharp disagreement about their relevance for HFTTR, where the dead fetus is not a research subject but a source of materials for research. There was strong resistance to Burtchaell's invocation of the Nazi analogy ([5], pp. 63–70; [28], pp. 690f.) on the grounds that this analogy is inapplicable to HFTTR and also "ethically repugnant" ([21], p. 27). Critics of the analogy with Nazi research stressed several morally relevant differences between the use of tissue from dead fetuses, following debatably immoral abortions, and the clearly immoral actions of the Nazi investigators in experimenting on living subjects against their will ([24], pp. 32–33; [21], pp. 27–28; [28], pp. 699–700). Perhaps the majority could have more effectively shown the inappropriateness of the Nazi analogy and the inapplicability of the Nuremberg principles to *the use of cadaveric fetal tissue in transplantation research following deliberate abortions.* A well-established alternative framework is the paradigm of and principles involved in cadaveric organ and tissue donation and transplantation, which are backed by a strong societal consensus [8]. Whether principles or case-judgments are central, interpretation is necessary, and the Panel failed to interpret the various principles and analogies as imaginatively and cogently as possible and desirable.[3]

Reductionist Explanations of Consensus on Panels and Commissions. In explaining the practical consensus among the members of the National Commission, MacIntyre noted its homogeneity, involving "upper-middle-class suburban Americans," and its political imperatives ([27], p. 612). However, Toulmin insists that the National Commission was more diverse than MacIntyre supposes, and that consensus would have developed even in the face of greater diversity. In a different way, critics of the HFTTR Panel have claimed that its consensus was foreordained because of the selection process, which resulted in the appointment of panelists who had previously supported HFTTR, who accepted abortion, or who had been beneficiaries of NIH. Thus, according to Burtchaell, "[t]here was no surprise whatsoever in the final vote of the panel" ([31], p. 67). By contrast, defenders of the Panel stress its fair and open-minded discussion ([31], pp. 68–69; [2], p. C5). Nevertheless, important questions arise about how representative a commission or panel must be of the diverse views held in the society at large.

Participation in the Panel's Deliberations. Did the panelists participate in the deliberations in good-faith? How was the consensual process used? Proponents and opponents of HFTTR probably viewed each other as failing to participate in good faith with sufficient openness to the other position. Burtchaell's comments about the Panel's composition and its operation "under something of a shadow" suggest this interpretation ([31], pp. 66, 69). In addition, at least some in the majority felt that the dissenters Bopp and Burtchaell failed to act in good faith when, without notice at the end of the second set of (and apparently final) meetings, after securing an agreement not to elaborate the brief "responses" to the ten questions, they deposited a long, elegant dissent, which would have overwhelmed the majority's report in length, quality of analysis, and strength of argumentation. Throughout the deliberations, the modest report had been crafted and, from the majority's standpoint, weakened in part to accommodate many of the minority concerns. However, if the majority had chosen to outvote the dissenters without serious efforts at both compromise and consensus in the formulation of the responses, the minority might have chosen to withdraw from the Panel and thus to de-legitimate the process and results. In short, there were grounds for a political decision to continue to participate in this way.

Panelists with prior strong moral convictions about HFTTR had some reason to worry that their participation in a process that resulted in consensus could help to legitimate the Panel's conclusion even if it differed from their own. Their participation could itself have become a form of complicity in evil. Of course, this risk was minimized because of the possibility of filing a dissent. In his own dissent, Daniel Robinson, an opponent of abortion who had been a very active and vigorous participant in the drafting of the report, indicated that he had tried to offer advice on public policy, within the context of the current laws permitting abortion and apart from "the question of the morality of abortion." However, his dissent registered his own "personal position" and "firm opposition to any form of Federal support for research making use of tissues obtained in this manner" since such research cannot redeem or exculpate abortion ([25], p. 73). And Bopp and Burtchaell ([5], p. 45) noted that "with the other panelists we have participated in the discussions and the drafting process, and have cast our votes for or against the various answers," but because of what they perceived to be serious inadequacies in the Panel's report, they dissented.

Consensus: Ethical or Political or Both? Was the consensus reached by the HFTTR Panel an ethical consensus? And what is the ethical relevance of the consensus? There was general consensus about what should or could be done as a matter of "acceptable public policy," but it is not clear whether this was ethical consensus in contrast to a normative consensus about public policy. Patricia King ([18], p. 250) notes that "[s]uccessful inclusion of ethical premises has also tended to foster the illusion that these bodies have achieved consensus at the level of ethical principle or even ethical analysis. In fact, the consensus usually comes at the level of practice and policy. Moreover, it is not clear whether an effort to reach consensus at the level of principle is either possible or desirable."

As previously noted, the HFTTR Panel agreed that federally-funded HFTTR, within certain safeguards, is "acceptable public policy." There was extensive and vigorous debate about the addition of the modifier "ethically," as in "ethically acceptable public policy," but the Panel finally settled for "acceptable public policy," in part because the chairman, a strong opponent of abortion in most cases, vigorously opposed the modifier ([29], p. 140 et passim; [31], p. 67). Without this concession, he probably would have dissented and thus reduced the political significance of the report.

However, in retrospect, this concession played into the hands of the critics of HFTTR, who could charge that the Panel abandoned ethical considerations altogether. Assistant Secretary for Health, James Mason

([19], p. 17), claims that the majority of the panelists indicated that "moral and ethical considerations were not central to their view of the issue." It is difficult to understand how he came to this conclusion. In fact, rather than denying the centrality of "moral and ethical considerations," the panelists in the majority simply offered a different view of the dictates of morality and ethics and a different balance of "moral and ethical considerations." However, such interpretations have led some panelists to believe that they should have pressed for stronger language in the report – for example, in holding that HFTTR is "ethically acceptable" as well as "acceptable public policy" – because the efforts to find compromise language to gain the support of more panelists left the report vulnerable at points and subject to neglect, misuse, and misquotation.

IV. THE HFTTR PANEL AND SOCIETAL CONSENSUS

How were the HFTTR Panel's deliberations related to an actual or prospective societal consensus? Did the Panel attempt to discover or create a societal consensus? Probably such a governmentally-appointed body would not be created without some concern about a lack of professional or societal consensus.

Presupposed Consensus. As already noted, the panelists often presupposed and sometimes even appealed to a pre-existing societal consensus, on at least some matters, such as the negative paradigm case of Nazi experimentation and the principles embedded in the Nuremberg code, but for the majority these did not dictate a particular conclusion on HFTTR with fetal tissue from deliberate abortions. Furthermore, in its argument for recognizing the woman who chooses an abortion as the appropriate decision-maker about donation, the Panel stressed the congruity between her role in this situation and other societal practices of organ and tissue donation, which apparently have a societal consensus. Other examples could be given. But, just as the National Commission did not try to "gauge public sentiment" for its recommendations ([27], p. 608), so the Panel did not focus on public sentiment, except briefly in some discussion about what the public believes about abortion ([28], pp. 660–663). The Advisory Committee ([2], p. 4) to the Director of NIH, to which the Panel's report was submitted, praised the Panel's consensus in part as a reflection of the societal consensus on this controversial topic. Beyond attempting to reflect a societal consensus, another way to illuminate a policy-maker's decision is to bring a range of perspectives and arguments to bear on the controversy and even to reconceive the problem. (I will return later to this possibility.) In retrospect, the critics of HFTTR on the Panel may have used their platform, particularly in their dissents, in more politically astute ways than those who supported HFTTR; after all, the position taken by the dissenters remains public policy.

Concerns about the Creation of New Societal Consensus. Even though the Panel did not explicitly appeal to an existing societal consensus about funding HFTTR, opponents worried that the provision of government funds and acceptance of the benefits of successful HFTTR would lead to a societal consensus that abortion is acceptable. Hence opponents were concerned about the symbolic legitimation of abortion through the indirect approval that would allegedly accompany the provision of federal funds for HFTTR using tissue from induced abortions. In his dissent, Rabbi Bleich ([4], p. 40) worried about the "aura of moral acceptability," because "federal funding conveys an unintended message of moral approval for every aspect of the research program." By contrast, the majority insisted that it is possible to support and accept benefits from HFTTR without approving of the abortions that produce the tissue, just as society now funds and accepts organ transplantation without approving of the homicides or accidents that make the organs available ([24], p. 35).

The Significance of International Consensus. So far my discussion has focused on the U.S. LeRoy Walters ([2], p. C5; [31], pp. 13–16), who chaired the ethics discussion of the HFTTR Panel, notes that the Panel's position, in contrast to DHHS's indefinite moratorium, is in accord with the international ethical consensus on this research, as reflected in the recommendations of various committees or deliberative bodies around the world. (At least nine such reports had appeared by December 1988 when the HFTTR Panel made its report and several more have appeared since then.) Despite important cultural differences in the countries involved, the "remarkable similarities" of these reports represent "an impressive international consensus on the ethical standards that should govern the use of fetal tissue for research" ([2], p. C5, Walters' Testimony). (Of course, questions may arise about whether the other reports reflect a societal consensus in their countries of origin.) According to

Walters ([31], p. 13), the "remarkable ethical consensus" in the fourteen reports available by April 1990 is that HFTTR is "ethically acceptable in principle" ([31], p. 13). Then they add conditions to ensure the ethical conduct, in practice, of such research, by limiting use to cadaver tissue and by attempting "to insulate the humanitarian use of fetal tissue from the abortion decision on the one hand and the commercial sphere on the other" ([31], p. 13). While a consensus, whether on a panel or within a country or around the world, does not guarantee that a position is "ethically correct," Walters contends that "we are less likely to make a serious moral mistake when numerous groups of conscientious men and women from around the world have sought to study the issue with great care and have reached virtually identical conclusions about appropriate public policy" ([2], p. C5).

Walters ([31], p. 15) further argues that "[t]he burden of proof on the ethics of fetal tissue research rests on DHHS officials who would have U.S. public policy differ from this impressive [international] ethical consensus." The political strength of the right-to-life movement in the U.S. is certainly a factor, but critics of HFTTR can also point to special conditions in the U.S. that might justify departing from the international consensus. First, U.S. abortion laws are less restrictive than those in much of Europe ([15], pp. 30–31). Thus, there may be more reason to fear the impact of HFTTR on abortion decisions and on the societal acceptance of abortion in the U.S. than in many other countries. Second, the U.S. has allowed more commercialization of and imposed less regulation on both abortion clinics and tissue procurement than some other countries. Furthermore, it has been suggested that the HFTTR Panel did not adequately appreciate the institutional pressures that exist in the U.S. ([3], pp. 1079–1082). By contrast, the U.K. Polkinghorne report [12] called for a national "intermediary" organization, with governmental funding, to separate the *practice* of abortion and the use of fetal tissue – a recommendation that goes beyond the U.S. panel's efforts to separate the *decisions* about abortion and donation. Thus one question in the U.S. is whether the limited consensus about what should be done if HFTTR goes forward, whether with private or public funds, can actually be implemented, in view of these societal and institutional factors. In view of the Panel's consensus, based on a variety of reasons, about separating abortion decisions and practices from the donation and use of fetal tissue in transplantation research, questions of feasibility become important for an ethical assessment of public policy. The chair of the HFTTR Panel, Judge Adams ([1], pp. 25–26), a strong opponent of abortion, finally supported federal funding of HFTTR largely to have a way to implement the "safeguards" endorsed by the Panel.

V. CONCLUSIONS

Consensus is a slippery concept, and its contemporary use often reflects unclarity about the extent of agreement, the nature of the agreement (e.g., what or why), and the process that generates the agreement. In addition, consensus cannot be taken as unquestionably valuable, either as an end or as a means, but requires a more nuanced evaluation.

While consensus can increase our confidence in particular and general judgments, any consensus may be mistaken (See [7], p. 385). Furthermore, the significance of the consensus reached in any group such as the HFTTR Panel depends in part on the fairness of the process, the range of information presented, the open-mindedness of the panelists, etc. It is also important not to overlook the psychosocial dynamics of group formation and interaction (see [20]). Even if a Panel's consensus has limited *ethical* significance, it may still have major *political* significance, particularly for public policy formation.

However, efforts to achieve consensus are often costly. One cost for the HFTTR Panel was the excessive amount of time spent trying to find formulations and words that would gain the agreement of more and more panelists. The carefully-crafted responses, worked out through the group process that consumed energy as well as large amounts of the limited available time, usually resulted in divided votes (but with only small minorities). Seeking this consensus may have diminished the intellectual and rhetorical quality of the final product. However, as I have suggested, any effort to short-circuit the process of consensus formation – perhaps by pressing for earlier votes on more sharply-worded formulations – could have produced a less effective product from a political standpoint, for panelists in the minority might have withdrawn in protest and thereby de-legitimated the Panel's report.

Clearly, then, here was a significant tradeoff, for the Panel's report with its brief "responses," followed by its brief "considerations," is not intellectually or rhetorically satisfactory. The report may fail to persuade because it does not clearly and fully explain the reasons that in collective deliberation could lead people with different views about the status of the fetus and about the moralitiy of abortion to support federal funding for HFTTR using tissue from deliberate abortions within certain "guidelines." "It was probably necessary," as King notes ([18], p. 251), "to describe the process that resulted in acceptance of this point rather than merely stating it." Perhaps a deeper problem, King ([18], pp. 251–252) continues, is that in its "drive to achieve consensus, the panel gave insufficient attention to diverse views, to raising new questions, to stimulating debate, and to furthering societal discussion of controversial matters. Perhaps consensus was achieved at the expense of other functions that these national bodies ought to perform."

Furthermore, the Panel sought consensus in response to the Assistant Secretary's definition of the problem through his ten questions, which severely constrained the discussion because of their focus on the linkage or connection of HFTTR with abortion. As King ([18], p. 252) argues, a more neutral set of questions – for example, "under what set of circumstances, if any, should the federal government support human fetal tissue transplantation research?" – would have allowed the exploration of a broader range of perspectives and frameworks. Nevertheless, the Panel could and perhaps should have redefined the problem in terms of the consistency or inconsistency of HFTTR with the paradigm case of and the principles undergirding the contemporary use of cadaveric tissue in education, transplantation, and research. This approach would have made it easier to explain why the Bopp-Burtchaell effort to use the Nazi analogy and the principles undergirding research with human subjects was beside the point. And the Panel could have addressed the ten questions in a new context.

Can such a governmentally-appointed panel, seeking consensus, function only in terms of what Daniel Callahan [6] has described as a *regulatory* role rather than a *prophetic* role? Can such a panel ever be prophetic in the sense of offering social criticism? Prophetic judgments may (1) appeal to transcendent norms by which to criticize and evaluate current social norms and practices, (2) offer a different interpretation of the meaning and ranking of current social norms, or (3) call for conformity to accepted norms. The first form of prophecy is not a likely prospect for a governmental panel, but the other two are possibilities. Reflecting the last two approaches, Michael Walzer ([34], p. 89) describes much Hebrew prophecy as "social criticism because it challenges the leaders, the conventions, the ritual practices of a particular society and because it does so in the name of values recognized and shared in that same society." Prophecy in the second and third senses presupposes some consensus, rather than only opposing consensus.

The prophetic approach may also sow the seeds of a different future consensus. Walter Harrelson ([16], p. 256) has observed that DHHS received "what it desired" but not "what it most needed" in the HFTTR Panel's report. What it most needed was "a rhetorically and aesthetically attractive report ... [with] a language and a set of images that will help a polarized community begin to build elements of consensus" ([16], p. 256). Such a report could have contributed more significantly to the societal conversation and formation of a consensus.

The extensive concurring statement, prepared by John Robertson and signed in whole or in part by ten other panelists, including myself, was intended to counterbalance the dissenting statement by James Bopp and James Burtchaell. However, it may not have fully achieved its goal, in part because some of its language and its conclusions provided materials that critics could use, even if inappropriately, to charge that the majority sacrificed ethical considerations for "significant medical goals" ([5], p. 70; [19], pp. 17–18).

It is probably not fair to judge the HFTTR Panel's report on the basis of its subsequent fate. However, DHHS, in effect, adopted the minority dissent rather than the majority report. At Congressional subcommittee hearings in April, 1990, Representative Henry Waxman (D-CA) challenged Assistant Secretary Mason about precedents for DHHS rejection of a *unanimous* advisory committee report – the Advisory Committee to the Director of NIH had unanimously endorsed the HFTTR Panel's majority report and recommendations ([31], p. 81). The Assistant Secretary conceded that such a rejection is rare. But would a stronger report by a majority of panelists, with a less general consensus, have changed public policy? The answer is very probably negative, but the societal conversation, with the possibility of a new future consensus, would certainly have been better served by a report that was intellectually and rhetorically richer.

University of Virginia
Charlottesville, Virginia
USA

NOTES

[1] Much of this section devoted to the case study is drawn from my essay, "Deliberations of the Human Fetal Tissue Transplantation Research Panel," in [11]. I have also incorporated elsewhere in this paper some other ideas and formulations from that essay and from "Ethics, Public Policy, and Human Fetal Tissue Transplantation Research" [10].

[2] These questions had been developed by a member of the Assistant Secretary's staff through an analysis of the literature and consultation with some academic bioethicists. See [11].

[3] When I note deficiencies in the work of the Panel, I do not blame any individuals and I certainly do not exempt myself from the criticisms. Constraints on the Panel's collective deliberations included the limited time, tight schedules, and limited staff, all of which may have reduced the Panel's effectiveness in contrast to some other national bodies.

BIBLIOGRAPHY

1. Adams, A.B.: 1988, 'Concurring Statement', in *Report of the Human Fetal Tissue Transplantation Research Panel*, National Institutes of Health, Bethesda, MD, Vol. I, pp. 25–26.
2. Advisory Committee to the Director, National Institutes of Health: 1988, *Human Fetal Tissue Transplantation Research*, National Institutes of Health, Bethesda, MD.
3. Annas, G. and Elias, S.: 1989, 'The Politics of Transplantation of Human Fetal Tissue', *New England Journal of Medicine* 320, 1079–1082.
4. Bleich, J.D.: 1988, 'Fetal Tissue Research and Public Policy', in *Report of the Human Fetal Tissue Transplantation Research Panel*, National Institutes of Health, Bethesda, MD, Vol. I, pp. 39–43.
5. Bopp, J. and Burtchaell, J.: 1988, 'Human Fetal Tissue Transplantation Research Panel: Statement of Dissent', in *Report of the Human Fetal Tissue Transplantation Research Panel*, National Institutes of Health, Bethesda, MD, Vol. I, pp. 45–71.
6. Callahan, Daniel: 1991, lecture and discussion on June 14 at the Hastings Center.
7. Caws, P.: 1991, 'Committees and Consensus: How Many Heads Are Better Than One?' *Journal of Medicine and Philosophy* 16, 375–391.
8. Childress, J.F.: 1989, 'Ethical Criteria for Procuring and Distributing Organs for Transplantation', in J.F. Blumstein and F. A. Sloan (eds.), *Organ Transplantation: Issues and Prospects*, Duke University Press, Durham, NC.
9. Childress, J.F.: 1990, 'Disassociation from Evil: The Case of Human Fetal Tissue Transplantation Research', in L. Hodges (ed.), *Social Responsibility: Business, Journalism, Law, Medicine* 16, 32–49.
10. Childress, J.F.: 1991a, 'Ethics, Public Policy, and Human Fetal Tissue Transplantation Research', *Kennedy Institute of Ethics Journal* 1, 93–121.

11. Childress, J.F: 1991b, 'Deliberations of the Human Fetal Tissue Transplantation Research Panel', in K. E. Hanna (ed.), *Biomedical Politics*, National Academy Press, Washington, D.C., pp. 215–248.
12. Committee to Review the Guidance on the Research Use of Fetuses and Fetal Material: 1989, *Review of the Guidance on the Research Use of Fetuses and Fetal Material*, Her Majesty's Stationery Office, London.
13. Engelhardt, H.T. and Caplan, A.L.: 1987, 'Patterns of Controversy and Closure: The Interplay of Knowledge, Values, and Political Forces', in H.T. Engelhardt and A.L. Caplan (eds.), *Scientific Controversies: Case Studies in the Resolution and Closure of Disputes in Science and Technology*, Cambridge University Press, Cambridge, Eng., pp. 1–23.
14. Fox, J.: 1988, 'Overview of Panel Meetings', in *Report of the Human Fetal Tissue Transplantation Research Panel*, National Institutes of Health, Bethesda, MD, Vol. II, Appendix A.
15. Glendon, M.A.: 1989, 'A World Without Roe: How Different Would It Be?' *Hastings Center Report* 19 (July/August), 30–31.
16. Harrelson, W.: 1991, 'Commentary', in K.E. Hanna (ed.), *Biomedical Politics*, National Academy Press, Washington, D.C., pp. 255–257.
17. Jonsen, A.R. and Toulmin, S.: 1988, *The Abuse of Casuistry*, University of California Press, Berkeley, CA.
18. King, P.A.: 1991, 'Commentary', in K.E. Hanna (ed.), *Biomedical Politics*, National Academy Press, Washington, D.C., pp. 249–254.
19. Mason, J.O.: 1990, 'Should the Fetal Tissue Research Ban Be Lifted?' *The Journal of NIH Research* 2, 17–18.
20. Moreno, J.D.: 1994, 'Consensus by Committee: Philosophical and Social Aspects of Ethics Committees', in this volume, pp. 145-162.
21. Moscona, A.A.: 1988, 'Concurring Statement', in *Report of the Human Fetal Tissue Transplantation Research Panel*, National Institutes of Health, Bethesda, MD, Vol.I, pp. 27–28.
22. Nolan, K.: 1988, 'Genug ist Genug: A Fetus is Not a Kidney', *Hastings Center Report* 18 (December), 13–19.
23. *Report of the Human Fetal Tissue Transplantation Research Panel*: 1988, II Vols., National Institutes of Health, Bethesda, MD [Unless otherwise indicated, all references are to Vol. I]
24. Robertson, J.A.: 1988, 'Concurring Statement', in *Report of the Human Fetal Tissue Transplantation Research Panel*, National Institutes of Health, Bethesda, MD, Vol. I, pp. 29–38.
25. Robinson, D.N.: 1988, 'Letter to Dr. Jay Moskowitz', in *Report of the Human Fetal Tissue Transplantation Research Panel*, National Institutes of Health, Bethesda, MD, Vol. I, p. 73.
26. Sullivan, L.W.: 1989, Letter to Dr. William F. Raub, Nov. 2.
27. Toulmin, S.E.: 1987, 'The National Commission on Human Experimentation: Procedures and Outcomes', in H.T. Engelhardt and A.L. Caplan (eds.), *Scientific Controversies: Case Studies in the Resolution and Closure of Disputes in Science and Technology*, Cambridge University Press, Cambridge, Eng., pp. 599–613.
28. *Transcript of the Meeting of the Human Fetal Tissue Transplantation Research Panel*: 1988a, Sept. 14–16, National Institutes of Health, Bethesda, MD.

29. *Transcript of the Meeting of the Human Fetal Tissue Transplantation Research Panel*: 1988b, Oct. 20–21, National Institutes of Health, Bethesda, MD.
30. *Transcript of the Meeting of the Human Fetal Tissue Transplantation Research Panel*: 1988c, Dec. 5, National Institutes of Health, Bethesda, MD.
31. U.S. Congress, House of Representatives, Committee on Energy and Commerce, Subcommittee on Health and the Environment: 1990, *Fetal Tissue Transplantation Research*, Hearing, 101st Congress, 2d sess., 2 April, 1990, Serial No. 101–135.
32. Vawter, D.E., et al.: 1990, *The Use of Human Fetal Tissue: Scientific, Ethical, and Policy Concerns*, Center for Biomedical Ethics, University of Minnesota, Minneapolis, MN.
33. Walters, L: 1988, 'Statement to the Advisory Committee to the Director, NIH', in Advisory Committee to the Director, NIH, *Human Fetal Tissue Transplantation Research*, National Institutes of Health, Bethesda, MD, p. C 5.
34. Walzer, M.: 1987, *Interpretation and Social Criticism*, Harvard University Press, Cambridge, MA.
35. Windom, R.E.: 1988, Memorandum on 'Fetal Tissues in Research' to Director, National Institutes of Health, March 22.
36. Wyngaarden, J.: 1987, Memorandum to Robert E. Windom on 'Approval to Perform Experimental Surgical Procedure at NIH Clinical Center – ACTION', October 23.

PETER WEINGART

CONSENSUS BY DEFAULT

The Transition from the Social Technology of Eugenics to the "Technological-FIX" of Human Genetics *

I. INTRODUCTION

The following is a study in the interplay between the development of a special kind of technology and the emergence of public acceptance and consensus over its use. The case is that of human reproduction. The technology in question, unlike others, involves concepts of human identity and ethics. Other technologies may do that in indirect ways by transforming cultural patterns, but they usually do so in unanticipated and undetected fashion. In the case of technologies which directly involve humans, these effects are much more immediate and may even be part of the explicit objectives, even though their long range consequences may be little understood. One would expect that new technologies of this kind are more controversial from the outset and have a higher probability of being rejected. Their eventual acceptance involves mechanisms by which they become adapted to the ethical context, but they also transform this context.

The separation of sexuality and reproduction became the overriding issue in the late 1800's and early 1900's with sexuality being left to the individual, but reproduction being considered a legitimate concern of the state. The objective was the realization of behavioral change in the realm of human sexual life according to a scientifically postulated goal – hereditary health – with direct or indirect socio-political means. In the subsequent development of eugenics during the first four decades of the century the priority of direct or indirect means was the topic of professional and public debate. Both strategies were being realized, and some of them are with us till this day. At about the middle of the century the social-technological orientation of eugenics was abandoned and reproductive behavior became directed through medicalized

Kurt Bayertz (ed.), The Concept of Moral Consensus, 189–206.

"technological-fixes". The thesis is that this shift was not the result of a conscious debate on the part of those who promote the technology or of the ethicists' arguments prevailing over the eugenicists' but rather a co-evolution of the system of socio-political values which are the context in which the new technology emerges.

The focus of interest here are the philosophies underlying the technologies and the features that could explain why they remained controversial until they became medical "fixes", in spite of the fact that their eugenic potential has increased tremendously. Two consecutive historical phases in the development of the 'technology of human reproduction' will be examined here: the first section deals with eugenics as justified by German eugenicists, the second focuses on the transition as performed by American human geneticists.

II. EUGENICS AS A SOCIAL TECHNOLOGY – THE ILL-FATED APPROACH

Two aspects of eugenic 'theories' are important with respect to their implications for social reforms: their value systems underlying the definition of racial fitness or good hereditary stock, as it became called later, and the diagnosis of degeneration. With respect to both, eugenicists held varying views which, to some extent, were due to the development of science. The specific contribution to the demographic debate, which made eugenics appear as a modern science, was the issue of the hereditary quality of the population. From the start concepts of social Darwinism and race theories competed with those based on the emerging theory of heredity. Thus, Alfred Ploetz, who coined the term "race-hygiene", still held a vague notion of the perfection of the "type", by which he meant the improvement of the "overall constitution with respect to the selective and social struggle" ([18], pp. 94, 118). W. Schallmayer, on the other hand, although opposed to race theories also favored a "unification of physical and intellectual racial fitness" ([21], p. 370). This view reflected the early impact of Mendelism, in that Schallmayer opposed views that the breeding of one class of traits could only be achieved at the expense of another. He, too, was influenced by social Darwinism and thus held that until the provision of personal records of heredity "the racial value of individuals or categories of individuals would have to be judged on the basis of the phenotype, i.e., the results of individual development and achievements". As this measure was only an approximation it could not, in Schallmayer's view, serve as a reference for direct influence in the reproductive relations of human societies. The goal of eugenics, thus, had to be the adjustment of the reproductive rate of different strata in the population to their respective social value ([20], p. 428). Schallmayer wanted his eugenics to be understood as social biology ("Gesellschaftsbiologie"), but eighteen years after the foundation of modern genetics, this was clearly more a biological social science.

Alfred Grotjahn, who published his "Hygiene of Human Reproduction" ("Hygiene der menschlichen Fortpflanzung") in 1926, represented the 'moderate' wing of German eugenics. For him the task of eugenics was to delineate the physically and intellectually "inferior" ("Minderwertige") and to assure that these two groups would contribute less to future generations than those who were average or superior ([8], p. 182). He rejected the Darwinist concept of fitness and suggested that, given the impossibility of an exact norm, practical eugenics would have to make due with "everyday experience". If this seemed arbitrary, he nevertheless wanted eugenics to be linked to the catalogue of known hereditary diseases and conceived as a branch of social hygiene. Although the social Darwinist deductions with their radical consequences receded to the background, his seemingly more scientific eugenics had both more far reaching and more constrained implications.

According to the state of hereditary pathology, it was believed that the more important hereditary diseases were recessive and that therefore members of the family were also carriers of the disease and had to be excluded from (further) reproduction. Although Grotjahn believed this conclusion to be politically unfeasible, it did enter the draft of the sterilization law of 1932. He opted for the limitation of eugenics to a "negative" function, i.e., to prevent hereditary diseases. The augmentation of talents and genius through reproduction seemed impossible, given the state of the science of heredity.

Grotjahn followed Fritz Lenz in committing eugenics to the science of heredity, but Lenz, probably the foremost scholar of human heredity in Germany at that time, leaned politically to the right while also retaining a social Darwinist perspective. Thus he focused on the conditions of selection. He differentiated between biological and social selection, the latter being dominated by a notion of fitness, for which occupational selection could serve as an exemplary indicator. It is hardly surprising that he considered class differences to be "to a large extent" biologically determined ([3], pp. 94, 97).

Besides the definition of hereditary quality, the diagnosis of the factors of its degeneration is the second essential referent of eugenic social technology. This diagnosis was not so much a medical but a social one. It gave the eugenic movement its sense of political mission, and its seeming urgency served to attract the attention of policymakers and public alike. The belief that degeneration was, in fact, taking place, was held by the social Darwinists against Darwin and was an expression of the cultural pessimism setting in shortly before the turn of the century.

Schallmayer, after initially having joined in the cries of warning, cautioned about the difficulties of objectively determining degeneration. Yet he had little doubt that at least physical fitness had decreased when compared with our ancestors in pre-historic times ([20], pp. 278, 280). As a "hard hereditarian", Schallmayer believed that only reproductive selection could improve the race.

For Grotjahn, the idea of degeneration assumed more distinct contours as he moved it in the direction of hereditary defects [1]([8], p. 14). Grotjahn's concept of degeneration represented an important step towards the "medicalization" of eugenics but was ambivalent with respect to the political implications. The link of the concept of degeneration to the science of heredity facilitated the conclusion that "roughly a whole third of the entire population does not meet the requirements which we have to set for flawless, fully robust and healthy indivuduals" ([8], p. 15).

Lenz differentiated between a value-free and a value-laden concept of degeneration. The former pertained to the genesis and diffusion of pathological traits, the latter to that of "otherwise undesirable traits". The major factor contributing to degeneration according to Lenz was if the conditions of selection assumed the characteristics of counterselection, and it was beyond doubt to him that this was the state of all nations of the occidental culture ([3], p. 11).

The eugenicists' concept of degeneration led to their diagnosis of its causes as well as to its remedies, i.e., the design of the technologies to avert degeneration. It is from this basis that the (social!) conditions of reproduction became the focus. All analytical energy was devoted to the elucidation and elaboration of selective, counterselective, or non-selective factors, from the supposedly too low average age at marriage of mothers to the selective consequences of poverty, from alcoholism to urbanization ([18], pp. 149, 183).

Schallmayer was the first eugenicist to introduce the systematic analysis of social conditions on the basis of the concept of selection, evaluating social institutions according to their eugenic or dysgenic functions. According to him cultural influences act on two levels of selection, determining the age of sexual maturity and the rate of reproduction and, thus, the contribution to the "gene pool", as it would be termed today. "Reproductive selection" assumed special importance as "vital selection" had been neutralized because of highly developed forms of cultural organization. From the logic of this approach it followed that next to the evaluation of social institutions according to their selective functions, the institutions connected with human reproduction became the target of potential state intervention. The low rate of reproduction on the part of racially superior women and the underlying reasons for this (such as their entering the labor market, the spread of birth control among the higher income groups, and the increasing age of women at marriage) became the subject of concern for demographers and eugenicists ([20], pp. 196–215, 235).

Whatever the focus of diagnosis of hereditary degeneration, it was believed that social conditions were always responsible, i.e., social institutions and thus behavioral patterns. The project of race hygienic practice, or a social biology, meant far-reaching reforms of society according to eugenic principles. Some elements of this scheme can already be found among philosophers and utopians long before Darwin, but among the Darwinian biologists reform began to assume the characteristics of a social technology, whose design owed as much to the scientistic and technocratic concept of society on the part of its engineers as it relied on smiliar views among its 'victims'. The technology aimed at securing the hereditary health of the population as a whole, focused on the reproductive behavior of the population, either directly and/or indirectly through manipulation of the conditions of selection.

Eugenic social technology began to take shape about 1903 with Schallmayer's major publication which contained practically all its future elements. Quantitative population policy was to be complemented by qualitative measures; human reproduction no longer was to be considered given but accessible and subject to political intervention. Tax reforms, especially income and inheritance taxes, were identified as important mechanisms to influence the differential birth rates of the upper and lower classes. Various schemes of easing the burden of parents with a greater number of children, either by adjusting income taxes or by the introduction of a state insurance, were put forth.

Eventually two measures survived the grandiose model building of the eugenic social engineers: the eugenically justified control of marriage, and sterilization of individuals with hereditary defects, primarily mental diseases. Both measures proved to be easier to implement than the more complex institutional reforms, because they could rely on the infrastructure of the public health system, or at least on the acceptance of the authoritative role of the state and the medical profession. (Sterilization, decided upon by the directors of psychiatric and mental hospitals, became a state enforced eugenic measure in Germany in 1934 after a draft of the law which provided for voluntary sterilization had already been completed in 1932. Health certificates as required for the issuance of marriage permits were put into law in 1935.)

It was Lenz who explicitly claimed an "essential relationship" ("Wesensverwandtschaft") between race hygiene and the "Fascist idea of the state". "While the liberal and, in essence, also the socialdemocratic ideas of the state are based on an individualistic Weltanschauung, fascism does not recognize the value of the individual. Its ultimate goal is eternal life, which is perpetuated through the chain of generations, and that means the race" ([3], p. 415). For that part of eugenics which remained linked to the concept of selection, fascism was a logical consequence, and its dependency on an autocratic political and value system could not have been stated in more clarity. The required restructuring of a large number of institutions, the orientation of society to one principle, hereditary health, and the need to overcome the anticipated opposition, made a centralist, authoritarian state seem to be the obvious ideal. It was a prerequisite of the eugenic social technology.

The alleged paradox that the return to an assumed pre-modern state of nature which guided the ideas of the eugenically ideal agricultural communities was connected with a scientistic, technocratic utopia leads one to overlook the fact that eugenics conceived of itself as a modern science. And, indeed, the diagnosis of the hereditary quality of the population based on the "new biology" and the expected future capacity to predict hereditary defects on a population scale as well as the causal impacts of social institutions on them could claim modernity. However, the conception of eugenics as a social technology proved to be both politically and scientifically conservative. By taking on major entrenched social institutions and values and relying on the powers of a state whose philosophy ran counter to the general trend, eugenics could not survive the brief, though consequential, interval of authoritarianism.

Today Lenz' comments on Ploetz' utopian vision that direct intervention into germ plasm would liberate mankind from the harsh consequences of selection appear like an ironic footnote to history. "One would have to have atomic tweezers with which one could grasp and exchange the single atoms of the hereditary material, in order to really control hereditary variations. But that I believe to be impossible for all time" ([3], p. 455). Lenz could not follow Ploetz' vision which was to become the guiding principle of the "technological fix" of molecular biology.

III. THE TRANSITION TO MEDICALIZATION

In contrast to the accounts given by protagonists of the field or by its historians, the eugenic creed was not abandoned overnight. Hitler's regime (and to some extent Stalin's support for Lysenko) did serve the useful purpose of providing a welcome scapegoat which could detract from past unfounded claims and the ethical failures on the part of the profession in propagating and implementing eugenics. It also served to create the remarkable image of a discontinuity between the older history of eugenics and the recent history of human genetics. In fact, though, the continuities are greater than often imagined.

The change that did take place was only in part the result of a moral awakening in view of the perversion of science. The dynamics of research and the expansionist tendencies of the profession remained intact. The crucial factor must be seen in a fundamental change of the human genetics paradigm and thus in the practical strategy of intervention. This process can be understood as a shift from the eugenic 'social technology', which tried to direct human reproductive behavior 'from outside', i.e., through interventions of social institutions, to 'selfdirection'. The latter relies on human genetics providing only the technical solutions and the relevant information as part of the generally accessible knowledge about a medically 'reasonable' conduct of life individuals can adhere to when making their reproductive decisions. The change occurred gradually and the arguments surrounding it reveal the goals and motives implied in the transition. The debates were most open in the US and are therefore taken as example.

When Herman J. Muller, first president of the newly founded American Society for Human Genetics, introduced the Society's journal in September 1949, he tried to distance the field of human genetics from eugenics by branding it as politically racist and scientifically mistaken. He saw the most important step to prevent repeating the mistakes of the past in a close link of human to general genetics. Biochemical methods for the identification of genes would inform medical doctors about the reality of hereditary processes and their importance for medical diagnosis, therapy and prophylaxis ([15], pp. 5, 7, 8). In the area of quantitative studies of heredity, one of the primary problems was to determine the relative frequency of genetic differences and their effects on specific traits in the population. It had to be pointed out, however, that there is no fundamental difference between the extreme but rare genetic differences interesting for medicine and the less extreme but more common ones interesting to the physical anthropologists. With this, Muller, a professed eugenicist himself, described the open border of human genetics where there is no unequivocal dividing line between positive and negative eugenics.

Muller explicitly addressed the issue of what position society should take with respect to eugenic topics. While research was to have priority, the application of the accumulated knowledge could not be ruled out. One type of application he saw in the improved control of somatic constitution where genetics would come to serve organisms without changing their genetic basis. The other area – and an open question – would be the direction of reproduction. Muller had no doubt that eugenics, "the social direction of human evolution", was a most "profound and important subject", but that "the heat and the misunderstandings of present political controversy, and the prejudices rampant in all existing societies, make very bad soil for the development of sound eugenic policies at the present time" ([15], p. 17).

Thus, Muller's programmatic statement opening the era of human genetics outlined the main elements of the scientific and professional strategies as well as the perspective of a "new" eugenics. While human genetics was to be established with a view to medicine to which it was to offer its services in the analysis of rare genetic defects that could be identified within the framework of a consensual concept of disease, Muller had not given up the hope for a truly human eugenics. It is indicative that he held on to this perspective with the same benevolent posture with which the eugenicists had pointed to the political unfeasibility of their plans for mandatory sterilization and eugenic marriage permits.

The crucial aspect of the integration of human genetics into medicine was a shift in focus from the genetic make-up of a population to that of individuals. This shift implied a concentration on rare genetic defects which could be identified as diseases in favor of the less focused and value-laden research on 'normal' traits. This shift was brought about by the advances in genetics and was supported by the moral and political discrediting of eugenics.

The same paradigmatic change toward 'medicalization' and the continued persistence of eugenic orientations can be observed with respect to the practice of human genetics. Lee Dice, who was eugenically oriented and directed the Heredity Clinic at the University of Michigan founded in 1940, sought to promote genetic counseling in his Presidential Address before the ASHG in 1951. According to Dice, a characteristic difference between the new practice and the older eugenics could be seen in voluntary sterilizations or in abstinence from reproduction. In a democracy any program for the improvement of human heredity had to be based "on the voluntary cooperation of the citizens". The precondition for this arrangement, however, was the expert advice from human geneticists, although genetic counseling was still an imperfect art. Dice expressed the emerging trust the geneticists had in the rationality of individuals when he said that cooperation of people in a program of voluntary limitation of reproduction of inherited defects was possible and that it "would be an abnormal person indeed who would not refrain from having children" if he or she was aware of bearing a high probability of transmitting serious defects ([6], p. 2). Although Dice did not want geneticists or government to assume the responsibility for deciding on sterilizations and although he also opted for non-directive counseling, he still thought in terms of the hereditary quality of the population as a whole as the referent for human genetics. Human geneticists were not only interested in "the decrease of harmful genes, but also in the increase of desirable ones," according to Dice, and he believed that progress could be made "in the discovery of the factors involved in the production of superior human traits when this problem is given the attention it deserves" ([6], p. 6). With this position he probably represents the generation of practitioners who made the transition from 'autocratic eugenics' to 'democratic human genetics', if one accepts that simplification [2] ([9], p. 253).

Fear for the quality of the gene pool was aroused by Muller in his 1949 address and became widespread again among human geneticists in the 1950's. Muller had diagnosed a rising load of mutations because the balance of new mutations and their elimination was disturbed by the effects of modern medicine and by increasing radiation in the environment. His somber picture of the future was a revival of the eugenic argument against the long term consequences of medical practice with an important difference: the dysgenic effects were devoid of all categories of class and race, and the aseptic language of genetics did not reveal any social value references. Politicization set in, nonetheless. Muller's thesis entered the discussions about the dangers of nuclear testing and his row with the population geneticist Theodosius Dobzhansky led genetics into the quagmire of the debate over atomic weapons and its anticommunist underpinnings ([16], pp. 111–176).

At the 2nd International Congress for Human Genetics in 1961, Muller initiated another round of the eugenic debate with his paper on 'Germinal Choice – A New Dimension in Genetic Therapy'. However, insofar as this paper presented for the first time the idea of circumventing social problems by way of a realistic purely 'technological fix' for the perceived eugenic problems, it also marked an important change in the debate. The idea itself, the use of artificial insemination, was not new, but the technology had become a eugenically relevant instrument because it had become possible to freeze male sperm for an indefinite period. If one agreed with Muller that degeneration was inevitable given the increasing load of harmful mutations, and that coercion must be excluded, this method seemed particularly suitable. The choice of genetic material, with this method, would be made according to the 'social value' of the spender. The choice to use this technology would remain with married couples. The pool from which they could choose would be virtually unlimited, since it could include spenders who had proven their 'value' during their lifetime.

Muller made explicit the logic of 'technological fix' on which his scheme relied. Given voluntary acceptance the method was particularly adapted to democratic society since it sidestepped the problems with which traditional methods were faced. Indeed, reliance on differential reproduction and the requisite political intervention was no longer necessary. "The notorious reluctance of the innately ill-endowed to admit their deficiencies and to limit accordingly the size of their families tends to lose its genetic importance in a population that is being renovated anyway" [17].

Discussions over Muller's explicitly eugenical vision of germinal choice are a good indicator of the eugenic potential still existing within the community. His position was by no means severely criticized but rather drew supportive comments from such leading scientists like J.B.S. Haldane, Frederick Osborne, Ernst Mayr, James F. Crow and Francis Crick. One of the few critics of Muller's "benevolent utopia" was the geneticist Leslie C. Dunn who pointed to the 'historicity' of the value judgments implied in any breeding concept and also warned the human geneticists of the still effective dangers of a politicized eugenics ([7], pp. 3, 11).

Nevertheless, a fundamental change in the character of the discipline and a strengthening of the orientation toward medical practice began in 1959 when British researchers and Lejeune in France decoded the chromosomal basis of three frequent abnormalities, the syndromes of Down, Klinefelter and Turner. For McKusick, in retrospect, it was the birth of clinical genetics and the fusion of human genetics and medical genetics. The result was the completion of a long process of medicalization which McKusick acknowledges explicitly: "Medicine has given focus, direction and purpose to human genetics" ([10], p. 271).

Adding to this process was the development of routinized screening programs for the discovery of genetic defects, first applied to the 'inborn errors of metabolism', phenylketonuria, then to Tay-Sachs and sickle cell anemia. (Massachussetts passed the first law providing for the mandatory PKU-screening of newborns in 1963.) While at first sight the screening programs seemed to fit unproblematically into the medical paradigm, it must be remembered that they also represented the fulfillment of an old eugenic dream, namely, to have a comprehensive data base on the hereditary health of a population in order to allow eugenic intervention. How thin the line of demarcation between medical and eugenic applications was, human geneticists experienced with the implementation of screening programs for Tay-Sachs and sickle cell anemia in the early 70's. It not only revealed the political naivete of the scientists who did not anticipate the political implications of their technology for different ethnic groups (Jews and Blacks to whom the programs were administered) but also a new ethical dilemma. Human genetics was now able to inform people about a defect (from the early 70's on even 'carriers' who did not show any phenotypical signs could be identified) without being able to provide a therapy. Crossing the borderline from therapeutic to preventive medicine was seen as a new challenge forcing practitioners to consider the "potential psychological and social impact" of the respective programs ([5], p. 573).

The year 1966 marks the beginning of the new era of prenatal diagnosis. The development of amniocentesis gave genetic counseling a powerful diagnostic technique and strengthened its function. At the same time it had opened a completely new range of options and responsibilities for future parents. To avoid the risk of defective offspring it was now no longer necessary to abstain from having children altogether. Prenatal diagnosis, for the first time, enabled human genetics to refrain from behavioral changes as a means of therapy and, in a sense, to technically 'evade' the disease, i.e. avoid its incidence without far-reaching changes of social behavior. However, amniocentesis did imply selective abortion and the ethical problems connected with it, and it also implied the continuation of genetic defects in all those cases where parents risked a new pregnancy and the fetus turned out to be a phenotypically healthy carrier of the parents' genetic defects. This posed to human geneticists once again the classic eugenic problem of modern medicine neutralizing selective mechanisms and thus contributing to a gradual degeneration of the 'gene pool'. In a way it tested how far the human geneticists had come in renouncing their claims to be authoritative wardens of the gene pool.

If at first sight it would seem that in the wake of advances in modern genetics, human genetics had concentrated increasingly on the therapy of diseases and thus retreated to the consensus encapsulating the medical concept of disease, this hides the fact that precisely these techniques drew this consensus into question again. While it was stabilized with respect to very severe defects and diseases, the new techniques increasingly allowed the diagnosis of unknown, rare and far less serious defects. Thus, although the span of eugenic visions had been limited by diagnostic techniques, it was widened again by the dynamics of their development. The ambivalence with respect to goals rapidly became an issue of conflict within genetics and human genetics. The old debate over potential positive-eugenic application began again. This debate, beginning in the US in the early 1970's, was fired by the specific issue of the XYY-Chromosome and its purported relation to criminality ([2], pp. 34, 36, 43; [4]; [22]).

The politicization of the XYY-debate affected the human genetics community and led to sharp conflicts inside the profession. The issues involved were the technical dynamics of cytogenetic diagnostic methods and their social and ethical implications, professional expansion and the newly emerging relationship of human genetics to medicine. Arno

Motulsky believed that the discipline had gained public respect because it had turned from the purely social concerns of early eugenics to the "entirely medically oriented preoccupation of recent decades ... However, history repeats itself, and concern with the social and public issues of human genetics is again appearing" ([11], p. 119). Additional trouble on the public front was stirred up by the debates over the heredity of intelligence and over the implications of sociobiology. It is indicative of the sensitivity of the human geneticists that Motulsky, while he could announce that the field had "become 'medicalized'", feared the spectre of the 'horrible misuse' of human genetics in the 1930's and warned that by diversifying it might lose ([13], pp. 125, 131).

Motulsky had repeatedly reflected on the troubled demarcation between 'medicalized' human genetics, and the eugenic implications of the new reproductive technologies as well as the related debates on genetic engineering, behavioral genetics and hereditability of intelligence. The crucial issue became the alternative between the orientation toward the gene pool or toward the individual. In a talk before the 4th International Conference on Birth Defects in 1973 in Vienna with the ominous title "Brave New World?", Motulsky gave an analysis of the ethical implications of all available techniques, starting out with the warning against rushing to their application under public pressure, since the regulation of human behavior, the genetic determination of normal traits as well as common diseases and birth defects, was largely unknown. His basic principle to leave decisions necessitated by new discoveries in the hands of the individual, even if this implied social costs and a deterioration of the gene pool, was justified politically and ethically. Motulsky's analysis could be read as a catalogue of principles of a 'democratic' human genetics. It essentially boiled down to a pragmatic attitude toward the new techniques, reflecting a consensus that they should be judged within the framework of the medicalized and individualized orientation of human genetics [12].

This development signifies the transition to a rationality of individual choice, which presupposes, however, that the underlying scientific categories, methods and value judgments all be accepted. The eventual abandoning of concerns for the frequency of defective genetic material in the population did not come suddenly nor has it been complete, but it was facilitated by scientific discoveries. Population geneticists had maintained as early as 1917 that the expected effects of sterilization in eradicating genetic defects were mistaken. In the late forties and in the fifties the debate over the 'load of mutations' due to the impact of radiation raised the issue of genetic degeneration again. But, though unresolved in principle, model computations of the increase in frequency of, for example, a phenotypically curable, recessive genetic disease like PKU showed that, under certain assumed conditions, it would take a hundred generations for the initial incidence to quadruple. Such time intervals extend far beyond any perspective of political planning and even the absolute numbers are not sufficient to support the panic which the eugenic prophets were able to spread and use for their purposes. Any selection against recessive defects can have only extremely slow effects and negative eugenics is almost helpless in trying to extinguish them. Given new discoveries that ca. 70% of the population are carriers of at least one of the known defective recessive traits and extrapolating recent increases of knowledge about them which leads to estimates that the chance for any one individual to be completely free of all defective traits is about 1/50, reduces eugenic strategies of excluding people from reproduction to the absurd. Eugenics' claim to control can no longer discriminate between the few diseased and the many healthy. In view of such evidence, human genetics has given up claims to a regulatory role in the shaping of the genepool.

With the advent of DNA-recombination techniques, the possibility of genetic manipulation came even closer. But the consensus apparent by then that this technique should be restricted to somatic therapy rather than be applied eugenically was strengthened. Still, the fears of abuse by authoritarian governments were always present and only suppressed with references to trust in the rationality of the profession and the public alike ([14], p. 135; [1], p. 402). The basis of this consensus was, no doubt, the medicalization of the diagnostic and reproduction techniques running parallel to the delegitimization of any eugenic value references. This was reiterated by the 1983 report of the 'President's Commission for the Study of Ethical Problems in Medicine and Biomedical and Behavioral Research' which stated unequivocally that the goals of a healthy gene pool or the reduction of health costs could not justify a compulsory screening program and that screening and genetic counseling were 'medical procedures' which can be chosen by the individual who wishes to obtain information as an aid in personal medical and reproductive decisions ([19], p. 6).

IV. EPILOGUE

The developments described here trace the almost complete reversal of what used to be eugenic strategies. While they could be implemented only under the threat of an authoritarian political regime, modern human genetics can rely on a fairly widespread though by no means complete consensus. How is this process to be explained?

The thesis advanced here is that it is not to be explained by a sudden change of heart on the part of the human geneticists, or an improvement of moral standards in the society at large. The dynamics of genetic research are unchanged as is the logic of expansion and the drive for control which rule the behavior of professions. The process appears to be a complex co-evolution of different factors. An initially important condition is the close link between eugenics and the National Socialist regime in Germany because identification of one with the other involved eugenics in the downfall and moral condemnation of that system. One could speculate whether the eugenic strategy of changing social institutions could have survived, had the authoritarian framework continued to exist in which it thrived. But the world of knowledge is undivided and only for brief periods of time allows for niches. So the progress of genetics moved the field into the direction of providing technical solutions which could be integrated into the medical paradigm. These technological 'fixes' were consonant with the value system that has become dominant as the democratic order in Western industrialized countries. The orientation to the individual and his or her choices had a 'de-politicizing' effect in that virtually no interference with existing value patterns and institutions is implied in its strategies, or so it seems at least. Freedom and privacy of sexual life, physical integrity, the value of the individual and the privacy of mate selection and marriage, which eugenics had challenged head on, all remain untouched. Human genetics relies on the self-regulation effects of individual choices, even if it does so sometimes with ambivalence and uneasiness. The proof that this strategy represents a consensus in society is the widespread and virtually unquestioned acceptance of genetic counseling and amniocentesis.

As safe and sound as this arrangement may seem, some characteristics of the "technological fix" philosophy inherent to it are disquieting. While the eugenic social technology approach failed because it challenged social institutions, the situation is now reversed in the sense that technologies are advanced with agnosticism toward (or ignorance about) their de-stabilizing effect on values and institutions. Thus, more or less unfounded speculation about these effects has to compete with, and is sacrificed for, the certitude of the immediate, positive effects of the use of these techniques.

Meanwhile, the progress in genetics leads that subject back towards the genetics of normal human traits. The explosive expansion of chromosomal research and biochemical genetics has increased the knowledge about the genetic structure of humans. But the more encompassing that knowledge, the more difficult it will be to derive criteria for deciding what is "normal" and what is "pathological". Where individual suffering can no longer serve as such a criterion, research is driving the field back into the realm of social value-laden ambivalence. It is an open question whether the curative paradigm of medicine can hold its own against the preventive potentials of modern genetics. The conflict between the two fundamentally different orientations seems pre-programmed, and if the latter one prevails, it undoubtedly will, by way of individual risk-assessment on the basis of the information provided, have profound effects on human behavior, and thus indirectly on social institutions. Is it a mere coincidence then, that human geneticists today envision a 'genetic passport' for every newborn containing a list of his genetic polymorphisms together with warnings of certain environmental hazards and advice for a healthy conduct of life, while in the past Galton and Ploetz thought about a eugenic certificate which adolescents would be given upon reaching sexual maturity, containing the permission to reproduce and specifying the class of available mates? Actually, the differences between the two visions are more instructive. While the eugenic certificate was to be issued by a governmental agency which would sanction the specified mandate, the genetic passport would no longer rely on state enforcement but rather on the 'fine-tuning' of self-regulated individual behavior.

Thus, the danger of abuse of the new reproductive technologies by a eugenically oriented conspiracy from above does not seem to be the real danger. The concern is more justified that the established consensus will erode because the individual demand for the reproductive technologies stimulates ever more research while at the same time the public's ability to reflect upon the value references disappears. If anything, that reflection would allow us to resist the realization of the eugenic utopias through our own behavior.

University of Bielefeld
Germany

NOTES

* Parts of this paper are taken from P. Weingart, J. Kroll, K. Bayertz: 1988, *Rasse, Blut und Gene – Geschichte der Eugenik und Rassenhygiene in Deutschland*, Suhrkamp, Frankfurt.

[1] Degeneration for him was "a physical or intellectual decline based on hereditary traits of offspring in comparison to ancestors considered as without defects or at least essentially without defects".

[2] Sheldon Reed, who directed the Dight Institute from 1947 to 1977 and coined the term 'genetic counseling' because allegedly genetic hygiene reminded him of toothpaste and deodorant, did not even see any difference between eugenics and human genetics.

BIBLIOGRAPHY

1. Anderson, W.F.: 1984, 'Prospects for Human Gene Therapy', *Science* 226, 401–409.
2. Ausubel, F., Beckwith, J. and Janssen, K.: 1974, 'The Politics of Genetic Engineering: Who Decides Who's Defective?', *Psychology Today* 8, 30–43.
3. Baur, E., Fischer, E. and Lenz, F.: 1931, *Grundriß der menschlichen Erblichkeitslehre und Rassenhygiene*, 3rd ed., vol. 2: *Menschliche Auslese und Rassenhygiene*, Lehmanns, München.
4. Beckwith, J.: 1976, 'Social and Political Uses of Genetics in the United States: Past and Present', *Annals of the New York Academy of Sciences*, 265, 46–56.
5. Childs, B. *et al.*: 1976, 'Tay-Sachs Screening: Motives for Participating and Knowledge of Genetics and Probability', *American Journal for Human Genetics* 28, 537–549.
6. Dice, L.R.: 1952, 'Heredity Clinics: Their Value for Public Service and for Research', *American Journal for Human Genetics* 4, 1–13.
7. Dunn, L.C.: 1962, 'Cross Currents in the History of Human Genetics', *American Journal for Human Genetics* 14, 1–13.
8. Grotjahn, A.: 1926, *Die Hygiene der menschlichen Fortpflanzung. Versuch einer praktischen Eugenik*, Berlin, Wien.
9. Kevles, D.J.: 1985, *In the Name of Eugenics*, A. Knopf, New York.
10. McKusick, V.A.: 1975, 'The Growth and Development of Human Genetics as a Clinical Discipline', *American Journal for Human Genetics* 27, 261–273.
11. Motulsky, A.G.: 1971, 'Human and Medical Genetics: A Scientific Discipline and An Expanding Horizon', *American Journal for Human Genetics* 23, 103–123.
12. Motulsky, A.G.: 1974, 'Brave New World?', *Science* 185, 653–662.
13. Motulsky, A.G.: 1978, 'Medical and Human Genetics 1977: Trends and Directions', *American Journal for Human Genetics* 30, 123–131.
14. Motulsky, A.G.: 1983, 'Impact of Genetic Manipulation on Society and Medicine', *Science* 219, 135–140.

15. Muller, H.J.: 1949, Progress and Prospects in Human Genetics, *American Journal for Human Genetics* 1, 1–18.
16. Muller, H.J.: 1950, 'Our Load of Mutations', *American Journal for Human Genetics* 2, 111–176.
17. Muller, H.J.: 1961, 'Germinal Choice – A New Dimension in Genetic Therapy', *Proceedings of the 2nd International Congress of Human Genetics*, September 6–12, Rome, 1968–1973.
18. Ploetz, A.: 1895, *Die Tüchtigkeit unserer Rasse und der Schutz der Schwachen*, S. Fischer, Berlin.
19. President's Commission for the Study of Ethical Problems in Medicine and Biomedical and Behavioral Research, Screening and Counseling for Genetic Conditions, 1983, USGPO, Washington.
20. Schallmayer, W.: 1903, *Vererbung und Auslese im Lebenslauf der Völker. Eine staatswissenschaftliche Studie auf Grund der neueren Biologie*, Fischer, Jena.
21. Schallmayer, W.: 1918, *Vererbung und Auslese. Grundriß der Gesellschaftsbiologie und der Lehre vom Rassedienst*, 3rd ed., Fischer, Jena.
22. Witkin, H.A. *et.al.*: 1976, 'Criminality in XYY and XXY Men', *Science* 193, 547–555.

SIMONE NOVAES

BEYOND CONSENSUS ABOUT PRINCIPLES: DECISION-MAKING BY A GENETICS ADVISORY BOARD IN REPRODUCTIVE MEDICINE*

I. INTRODUCTION

Uncertainty about what constitutes a good (if not the best) medical line of conduct in a given situation is a characteristic feature of all medical practice. The physician is routinely confronted with limits in current medical knowledge, his own imperfect mastery of available knowledge, and the difficulty in distinguishing between his own limitations and that of the art when treating a particular case ([3]; [4]). Moreover, the increasing importance of research in hospital settings and of innovative technical approaches to routine medical problems has brought to the foreground the moral dilemmas underlying many situations of medical uncertainty. Not only do the usual normative references for dealing with patients no longer seem adequate under certain circumstances, but some new practices question fundamental moral and social values which previously seemed immutable ([5]; [6]; [7]; [8]; [11]); [12]).

One way of dealing with these problems has been to submit them to collective multidisciplinary bodies for discussion and advice. Many physicians are nevertheless reticent about having their experimental practices monitored by this type of procedure, which in their opinion slows progress on new therapeutic approaches to disease and disturbs the confidence which a patient should place in the attending physician. It is however not uncommon for physicians themselves, when faced with difficult or unusual cases, to resort to a collective form of decision-making for solving problems which are usually left to the individual physician's own judgment. There is an attempt to reduce, by consensus among peers, uncertainty as to the best way to act.

Kurt Bayertz (ed.), The Concept of Moral Consensus, 207–221.

This paper is about one such experience in France involving physicians engaged in one of the oldest forms of reproductive technology, artificial insemination with donor sperm (AID). Faced with the complexity of routine decisions concerning the genetic aspects of artificial insemination, notably controlling the donor's family history for serious hereditary disease and examining the validity of certain genetic indications for AID, they created a genetics advisory board to help them make what at first seemed essentially technical decisions. In fact, as the Board's work progressed, they discovered that, underlying uncertainty about these technical decisions and disagreements as to the best way to proceed, were often questions concerning the extent and the limits of a physician's responsibility with regard to his patients, and thus his right (or duty) to intervene – or to refuse to intervene.

How deep a consensus on moral issues can one expect in the proceedings of such a group? Jonathan Moreno raises this question in his paper [10], and introduces a distinction between consensus about particular questions, consensus about principles, and consensus about moral theory. Using Stephen Toulmin's account of his experience as a member of a national commission setting standards for experimentation on human subjects, he argues that it is easier for a group to reach an agreement on specific policies than it would be for members to agree on the moral reasoning behind their final decision. In other words, consensus about particular questions does not necessarily reflect consensus about principles or about a general moral theory guiding the members' decisions.

To a certain extent, his point does apply to decision-making by the genetics advisory board that we will study in this paper. However, the inverse relationship does not necessarily hold. As we will see, although consensus about principles does generally reflect strong moral cohesion among group members, it does not necessarily guarantee consensus on a practical level. First of all, even when a professional group sustains a common set of principles as governing their activity, these can be interpreted in various ways, when applied to the specific circumstances of a case. Principles can also conflict when a group is discussing a particular case, and group members will then have to agree as to which principles will be given priority in this instance. Moreover, as a group discusses and tries to work through a difficult case, it will sometimes be faced with the task of reassessing and redefining the situation; this eventually leads to a questioning of the principles which had originally served as guidelines, but which may no longer seem adequate in solving the problem at hand. The principles at stake in a particular situation may eventually be more general moral principles than those directly related to professional practice.

Thus, consensus about principles governing a common professional activity does not preclude controversy over the resolution of specific cases. On the contrary, I would like to show that such controversy is not only inevitable but ultimately useful, as it tends to reveal the wider scope of moral values effectively influencing the group members' collective decisions,[1] while contributing to a reconsideration of the principles guiding their professional activity as well as to a clearer understanding of the basic ethical dilemmas inherent to routine practice. Consensus about professional principles can provide a formal framework, within which diverging moral views may coexist and express themselves in the context of a commonly-accepted procedure for resolving conflict over concrete problems; however, this does not guarantee that acceptable solutions will be found to all the ethical dilemmas raised.[2]

II. THE FRENCH FEDERATION OF CECOS BANKS

The French Federation of CECOS semen banks is a network of 20 autonomous non-profit semen banks – the first of which came into existence in 1973. Concerned about the standards of practice of artificial insemination in France, handled for decades essentially by private practitioners during office consultations, the physician who set up the first bank chose a public university hospital setting: he perceived this as a means of controlling the technical quality of medical care, as well as a solution to his concern about frequent financial exploitation of distressed infertile patients in the private practice setting. New banks wishing to join the network often organized their services along the same lines and, in any case, had to agree to function according to a common set of principles when dealing with the problems raised by artificial insemination with donor semen.

These principles, which govern CECOS practice and which provide the basis for the consensus uniting the member banks into one federation, are derived from an ethical model of the ideal relationship between the diverse protagonists, condensed in a maxim "a gift from one couple to another" ("le don de couple à couple"). The relationship between donors and recipients is postulated as founded on empathy and solidarity; consequently, semen donation should be a free gift – an anonymous and a gratuitous act – from a fertile couple who has children to an infertile couple who wishes to have children. CECOS physicians define their task as mediators in the donor-recipient relationship, who protect *donor anonymity* and *recipient secrecy* regarding recourse to artificial insemination (which the latter often desire), while guaranteeing the *professional quality of the medical attention* both parties will be receiving. Moreover, as physicians, they do not wish to be responsible for social innovation in family relationships; they thus strictly limit their practice of artificial insemination to *medically-justified situations of infertility*.

This conception of their practice has resulted in a policy with two main characteristics relevant to this paper. First, as most semen banks, CECOS banks handle donor recruitment and screening but, unlike most banks, they also control access to donor insemination: AID with CECOS frozen semen is available only to heterosexual couples in which the husband has a medically-proven infertility problem or an important risk of transmitting a serious hereditary condition. Second, as is the case with most banks, semen donors must agree to remain anonymous; however, they are *not* paid, and only married men (or men established in long-term relationships) who have children and whose wives assent to semen donation are retained as potential donors. The aim of this policy is to improve the public image and moral standing of donor insemination, thus helping the donors and recipients themselves overcome doubts about the social acceptability of such a transaction. One of the main consequences of this policy, however, has been a scarcity of donors for the banks.

It is also important to note that physicians in France involved in semen banking for artificial insemination are working in what some lawyers have called a "legal vacuum", that is, no legal constraints have yet been established on reproductive technology or the various kinship issues which it raises. In this context, by creating a federation and setting up common normative references and practice standards, semen banks are protecting their activity from the effects of an unpredictable social and political environment.

III. GENETICS AND AID

In most semen banks in France, the genetic screening of donors involves doing a caryotype and taking the donor's medical history. With time and experience, however, CECOS physicians realized that such screening raised more problems than were first imagined. The medical history has proven to be the most important part of the screening (few donors are excluded because of caryotype abnormalities). As for the quality of the information obtained and the accuracy of the decisions made on the basis of that information, they seem to depend on the length of the interview with the donor and on the competence of the interviewer [14].

CECOS semen banks also receive a certain number of requests (from 1 to 2% of overall demand) from couples who have genetic indications for donor insemination (risk of transmission by the husband of a serious genetically-determined condition). Prenatal diagnosis is helpful, but some conditions cannot be detected by existing techniques; using donor semen can therefore be considered a valid alternative if the couple wishes to have children. However, given constant progress in prenatal diagnostic techniques, the criteria for genetic indications constantly need to be reevaluated.

Because of these problems with screening and with genetic indications for AID, CECOS banks began to consult clinical geneticists and were led to create what eventually became their Genetics Advisory Board. Its initial task was to elaborate a set of guidelines and practical recommendations for screening donors and examining the validity of genetic indications for donor insemination. But as CECOS physicians tested these guidelines by confronting the Genetics Advisory Board with specific cases from their daily practice, the Board came to realize that the setting of practice standards could not be handled from a purely theoretical perspective. It subsequently changed its mode of functioning and adopted a new approach based on case histories: difficult or unusual cases were submitted by the 20 banks to the Board for an opinion; the Board recommended a course of action and, at the same time, when necessary, revised its own guidelines.

As the Board's work has progressed, fundamental questions have arisen concerning the extent and the limits of legitimate recourse to clinical genetics in the context of reproductive medicine. These questions have come up, not because of an *a priori* stance favorable to ethical questioning: the Genetics Advisory Board is considered a technical committee and problems perceived as essentially moral are relegated to another committee, the CECOS Ethics and Deontology Committee, composed essentially of CECOS physicians and psychologists, and created by the Federation to discuss and propose solutions to the diverse non-medical problems raised by the application of CECOS principles in daily practice. Nevertheless, ethical dilemmas which could not be dissociated from the problems being tackled have arisen, mainly – we will see – as an unexpected consequence of a refinement in the genetic screening of donors. Confrontation between physicians specialized in the field of genetics but not always directly involved in the practice of AID, and the semen bank physicians, usually not trained in genetics but directly in touch with the concrete problems of routine practice, has proved indispensable in identifying the relevant moral issues underlying theoretical and technical options in genetic screening for reproductive procedures.

The Board's work on guidelines for genetic screening has ultimately involved the definition of an ethical stance regarding the use of clinical genetics in the context of reproductive medicine. For in contrast to genetic counseling, which raises some of the same questions, physicians engaged in reproductive technology control access to the essential resources which permit conception – in the case of AID, donor semen. Genetic screening thus gives a physician discretionary powers which he/she must decide how to use and for which ultimately he/she must account.

This aspect of work on guidelines has usually come up in Board discussions as an attempt to define the extent and the limits of their medical responsibility. This requires that the principle of ensuring the quality of medical attention provided both to donors and recipients, under the conditions of donor anonymity and recipient secrecy regarding recourse to AID, be translated into consensus about concrete technical choices. One of the Board's basic concerns here has been distinguishing between choices which represent a medical approach to disease prevention, in accordance with a physician's basic ethical stance of beneficent action in caring for his patient, and choices which could be interpreted as reflecting a eugenic policy, the main objective of which is an improvement, in the long term, of the human species through the control of hereditary factors in reproduction. This has sometimes involved weighting the physician's responsibility for the safety and the favorable outcome of a medical procedure, against the patients' right to autonomy in making reproductive choices. This ethical dilemma is complicated by the fact that medical concern with preventive action regards an as yet non-existent third party, the unconceived child.

IV. THE GENETICS ADVISORY BOARD

The Genetics Advisory Board is composed of three types of members: 1. CECOS physicians (clinicians and biologists) who submit the cases and ask the questions; 2. young clinical geneticists, who work with the CECOS banks, most of whom have training in pediatrics; and 3. clinical geneticists and cytogeneticists, who do not work in CECOS banks, but who are invited as consultants in view of their specialty. The Board meets twice a year and reviews approximately thirty cases per session. The cases are organized into three categories: 1. donors (semen and a few ovum donors, as the Board is sometimes consulted by a few teams doing in vitro fertilization); 2. genetic indications for donor insemination (and a few cases of genetic indications for ovum donation through in vitro fertilization); and 3. a new and unexpected category – *contraindication to donor insemination* (the infertile man's wife presents a major risk of transmitting a genetically-determined condition) – which has arisen as the consequence of a new mode of classification for screening donors.

General problems are usually discussed in relation to the individual case in which they arise; but recently the Genetics Advisory Board organized a special joint meeting with the CECOS Ethics and Deontology Committee, as the geneticists came to realize that ethical dilemmas could not always be treated separately from technical problems. The purpose of this first joint meeting was to reflect on three particular types of problems which come up frequently during Board discussions and are usually most difficult to decide. One of these problems is precisely that of contraindication to donor insemination. This special type of meeting also seems to respond to a new need: that of reviewing and critically assessing the Board's own activity and decisions.

At the first meeting of the Genetics Advisory Board in 1983, a list of genetic factors justifying the exclusion of a donor candidate was established (essentially, major chromosomal alterations, dominant disabling pathology, and some frequent serious recessive conditions which might therefore readily be found among recipients). In fact, three rather than two categories of donor candidates were created: 1. those excluded because of a serious risk of transmitting a genetically-determined condition; 2. those accepted without reservation (there is no apparent risk of transmitting a major handicap); and 3. those accepted with a *cumulative risk factor* (CRF), that is, donors whose family history indicates a risk of transmitting a genetically-determined condition, which however is either not serious or very rare, and can be prevented by attributing that donor's semen to recipients whose history does not present the same factor.[3] The creation of this last category of donors, which allows for fewer exclusions, was both an attempt to avoid the slippery slope toward excessive selection in the screening of donors, but also a response to concern with the useless exclusion of the scarce benevolent donor sought by CECOS banks. However, this new category requires at least limited screening of the recipient. The problems raised by selection, which all genetic screening implies, are thus displaced from control of what semen will be retained for preservation to decisions on how semen will be assigned for insemination.

This new genetic emphasis on selecting semen for a particular recipient (which is now added to criteria concerning the morphological characteristics of the couple – usually skin color, blood group, and sometimes hair and eye color) highlights the physician's role and responsibility as an intermediary in momentarily associating for reproductive purposes two persons who nevertheless are not to be recognized socially as a parental couple: the donor remains anonymous and recipients usually keep recourse to artificial insemination secret. The criteria and purpose of such matching must therefore be made clear to all the protagonists involved – and ultimately to society, if AID is to be considered a morally acceptable procedure.

When a physician is thus acting as intermediary between two anonymous parties in the non-sexual transmission of gametes for reproductive purposes, to what extent does his intervention make him responsible for the outcome of the procedure? In more general terms, to what extent should the practice of artificial insemination be guided by the endeavor to prevent hereditary disease in AID offspring? This problem often comes up in Board discussions of guidelines and decision-making, probably because the responsibility of a physician in the context of reproductive medicine has nowhere been precisely defined. It usually appears either as a preoccupation with legally-defined medical responsibility or as a broader concern with moral responsibility. During the discussion of difficult cases, in particular, an attempt is usually made to establish a working definition of this responsibility for CECOS physicians, to which their decisions can be referred.

Defining medical responsibility in the context of reproductive technology involves three distinct problems: 1. the screening of donors; 2. the risk of transmission by the recipient of a serious hereditary condition; 3. the risk involved in the use of prenatal diagnosis for an AID pregnancy.

With regard to donor screening, the intermediary is intervening as a physician, so he is responsible for applying the criteria of his profession in avoiding the deliberate transmission of disease. However, in the case of genetic screening, even an apparently healthy donor with healthy children may turn out to be a heterozygote for a recessive condition.[4] Eliminating all risk is therefore an illusory objective. Moreover, as progress is made in developing tests to detect heterozygotes for recessive conditions, fewer candidates will be acceptable for semen donation, if the objective of genetic screening is perceived as the elimination of all risk of transmitting hereditary disease. The problem must therefore be redefined: *how much risk-taking is acceptable* in the context of reproductive medicine?

This question has been raised in varied cases submitted to the Board. Should a man who has been cured of cancer be accepted as a donor? In other words, do hereditary factors contribute to causing cancer and what are the possible effects of chemotherapy and radiation therapy on such a man's descendants? Should men who are known heterozygotes for a genetic condition be accepted as donors? If not, why not? If so, under what conditions? Should there be systematic testing of donors at risk for frequent and serious recessive conditions? Given constant progress in testing for recessive diseases, how far should Board recommendations go in screening donors? The Board's work has led to the conclusion that no physician (or the institution supporting him) can guarantee the birth of a healthy child through the use of a medical procedure; he can only attempt to maintain risks at a limit *defined as acceptable* for that procedure by the profession (and ratified by society).

However, the task of defining such a limit in concrete terms is often a complex and controversial process. It is often quite easy for physicians to agree on the objective data which characterize a particular genetic disease (age at which the disease usually appears, symptoms, prognosis, statistical frequency of the disorder in the population, etc.) as well as to agree on the objective data which characterize the situation of the donor whose case is being considered (determining whether or not the donor is a carrier of a particular disorder, calculating the statistical risk that the disorder will be transmitted, etc.). But evaluating the *seriousness* of an hereditary condition (is congenital deafness a serious disorder?), the *importance of the risk* involved (is 5 or 10 percent a high or low risk?), and consequently deciding whether or not that condition justifies exclusion of a donor requires subjective appraisal of objective medical data. The weighting of CECOS principles which occasionally conflict also intervenes: for example, given CECOS concern with avoiding choices which could reflect a trend towards a eugenic policy, but also with the recipient couple's eventual desire for secrecy regarding donor insemination, should a donor at risk for transmitting a minor but visible defect (i.e., a harelip or a cleft palate), which could not have been inherited from the future parents, be excluded? At this point, the varied clinical experience of the Board members with their patients as well as their divergent moral views come into play, making consensus as to what constitutes an "acceptable" risk, in some cases, an unattainable ideal.

Is the physician responsible in the same way for deciding how much risk-taking is acceptable when it is the recipient of a medical reproductive procedure who is at risk for transmitting a serious hereditary condition? We must remember that, given the new category of donors with cumulative risk factors, recipients are screened (though not as thoroughly as are donors), so as to avoid inseminating the recipient with semen, which would result in the accumulation of the same risk factor. Such screening has occasionally led to the discovery that the recipient could transmit a serious genetically-determined condition to her offspring: it is usually dominant pathology, which the recipient has in a minor form but which the child could inherit in a major form, but it can also be a serious recessive condition linked to the X-chromosome. Is a physician justified in refusing donor insemination to an infertile couple for this reason?

The Board has usually based its working definition of medical responsibility (and consequent obligation to intervene in influencing the outcome of a procedure) on the distinction between reproductive technology and sexuality: only intervention in the latter, which implies no medical supervision, would be illegitimate. Is this distinction still relevant here? If so, where is the dividing line between the physician's responsibility for supervising the medical aspects and controlling the technical quality of a reproductive procedure and the recipient couple's autonomy in decision-making concerning a child for which they will ultimately assume parental responsibility? Can an acceptable risk be defined in this type of situation and in what terms?

These same questions have been raised in cases where recourse to prenatal diagnostic techniques could be a valid option for monitoring certain pregnancies. CECOS physicians are in fact often reticent to use (rare) benevolent donor semen to start a pregnancy in which there is a certain probability that the child will not come to term or that the pregnancy will have to be terminated by a therapeutic abortion. Nevertheless, who assumes the risk of such an outcome: the physician, the recipient couple, or the woman?

The idea of an acceptable risk has been even harder to define concretely in these last two types of situations than in the case of donor screening. Nevertheless, CECOS physicians feel that, because they are called upon to intervene in these situations as physicians, they must assume some responsibility for the risk of a serious hereditary disorder that recourse to a reproductive procedure might inflict on the child to be born. In attempting to justify this attitude, a new normative reference has come up in Board discussions: accountability of CECOS physicians to their benevolent donors for the appropriate use of their semen. This argument is, however, inconsistent with the fact that donors have given their semen anonymously and agree not to know anything about the outcome of donation; physicians are therefore, in principle, free to make all relevant medical decisions. By introducing the idea of accountability to donors, physicians appear to be redefining their responsibility in reproductive situations, in terms which are no longer merely professional: as trustees of the donor's semen, they invoke the donor's personal values with regards to reproduction, apparently as a means of justifying the intrusion of a broader range of moral values in deciding how semen is to be used.

The use of this type of argument to justify the medical veto of a reproductive procedure has never appeared totally satisfying to Board members, and the Board is still struggling to define an appropriate line of conduct for cases in which donor insemination appears to be contraindicated for genetic reasons. It does, however, indicate that deep consensus about the principles guiding a physician's professional activity is not sufficient for solving problems related to a medical procedure, in particular, one in which the outcome is the birth of a child. For ultimately, controversy over the best line of conduct and over the very definition of the extent and the limits of medical responsibility revolves around an undecidable dilemma: who, of the physician or the couple, have the best interests of the unborn child in mind, when deciding on a course of action? The difficulty in arriving at consensus about the best way to act seems to stem from the fact that medical and lay definitions of an acceptable risk, which often differ and frequently pit patients against physicians, also tend to overlap and eventually to conflict even in the physician's own reasoning, because the physician has other values from his private life influencing his professional attitudes, which in this case concerns the very sensitive problem of reproductive choices.[5]

V. CONCLUSION

Reproduction is today defined as an area within the limits of a physician's professional competence, but it must be remembered that the presence of (usually male) physicians at childbirth dates back only two centuries – a presence that was first justified by their monopoly of a new instrument, the forceps, designed to facilitate difficult births. New techniques in the area of obstetrics and gynecology, such as contraception, abortion, the monitoring of fetal growth or testing for fetal abnormalities during pregnancy are also recent acquisitions for the profession; nevertheless, most of these techniques have evoked much controversy over their legitimate use.[6] Any kind of medical intervention in reproduction thus obviously involves not only discussion of the normative principles guiding medical activity, but also of fundamental social values about sexuality and the family. Normative and value references which will usually seem adequate in discussing routine medical situations may thus appear insufficient when consensus must be achieved about the extent and the limits of a physician's involvement in a couple's decision to bear children. These new situations often appeal to a broader scope of values, usually embodied in personal moral convictions regarding the "good" life.

This does not mean that medical innovations in other areas do not entail controversy over fundamental social values. Quite to the contrary – and the abundant literature on bioethics testifies to this fact – many new therapeutic innovations (organ transplantation, radiation and chemotherapy for cancer, life-sustaining techniques, or even less dramatic, preventive aspects of medicine, such as vaccination) imply approaches to a patient's body, which raise questions about the extent to which a physician can legitimately intervene, even to save his patient's life. Routine medical practices also imply an underlying moral and social evaluation of the situation in which the physician is called upon to intervene; but this becomes evident only when controversy over the legitimacy of such intervention brings the underlying evaluation to the foreground.

Consensus on guidelines which define acceptable risk-taking in the context of reproductive technology thus necessarily involves working with values which may seem foreign to a medical situation (but which never are); in fact, it involves reassessing and redefining what is morally acceptable in terms of reproductive choices within a given society. This means that consensus about particular questions will be difficult to come by, even in a group composed exclusively of a professional group, whose activity is guided by the same ethical principles; their values regarding reproductive choices are influenced by many factors, all of which are not necessarily directly related to their professional training, experience or principles. The experience of the CECOS Genetics Advisory Board therefore also suggests that striving for consensus will ultimately involve redefining an ethical stance for physicians construed in terms of shared responsibility with the other protagonists for the outcome of a reproductive procedure. For once the diversity in the origins of values instructing a medical decision is recognized, it becomes impossible to exclude lay viewpoints from such decisions.

Centre de Sociologie de l'Ethique
(Centre National de la Recherche Scientifique)
Paris, France

NOTES

* This paper is based on my experience during the last four years as a non-participant observer at the working sessions of the CECOS Federation Genetics Advisory Board, as part of a research project on the banking of gametes for reproductive procedures, financed by the Mission Interministérielle Recherche Experimentation, France [11]. I wish to thank Kurt Bayertz, Raymonde Courtas, and Gwen Terrenoire for their comments and helpful suggestions on the first draft of this paper.

1 My work on the CECOS Genetics Advisory Board indicates that these moral values

tend to reflect differences in the clinical experience of group members, as well as the varied social composition of their professional group. This point cannot be treated extensively here, given the limits of this paper, but also because some of these aspects are still under investigation. For more detail, see [11]; [13].

[2] This is my way of rendering Kurt Bayertz's argument, on page 50 of his paper [1], in favor of ethical theories of consensus, which make dissent with regard to what is "good" as legitimate as consensus with regard to what is "just". Nevertheless, as he himself argues in the conclusion of his paper, dissent over what is "good" will necessarily lead in certain cases to dissent over what is "just". Attitudes concerning the value of fetal life and the ethicality of abortion might here be a typical example.

[3] The most common examples of cumulative risk factors are allergies and cardiovascular disease. For more detail on these guidelines for the genetic screening of donors, see [9].

[4] A recessive condition will usually appear in children of two partners who are both heterozygotes for the same condition; however, because they are themselves unaffected by the condition, they may be unaware of the fact that they are carriers of the gene until they have children. A donor, whose wife is not a heterozygote for the same condition, may be paired off with a recipient who is.

[5] An interesting international comparative study, directed by Dorothy Wertz and John Fletcher [15], of the attitudes of clinical geneticists from 19 different countries confronted with a typical set of difficult cases, shows how values concerning medical decisions related to genetics vary from one society to another, as well as among individual practitioners, in particular according to their age, sex, and religious practice.

[6] Not only the techniques, but also, more generally, the basic and applied research in the reproductive sciences have been the subject of much controversy since the beginning of the century. See [2].

BIBLIOGRAPHY

1. Bayertz, K.: 1994, 'The Concept of Moral Consensus. Philosophical Reflections', in this volume, pp. 41–57.
2. Clarke, A.E.: 1990, 'Controversy and the Development of Reproductive Sciences', *Social Problems*, 37 (1), 18–37.
3. Fox, R.C.: 1957, 'Training for Uncertainty', R.K. Merton, G.G. Reader, and P.L. Kendall (eds.), *The Student Physician*, Harvard University Press, Cambridge, Massachusetts, pp.207–241.
4. Fox, R.C.: 1979, 'The autopsy: Its Place in the Attitude-learning of the Second-year Medical Students', *Essays in Medical Sociology: Journeys into the Field*, John Wiley and Sons, New York, pp. 51–77.
5. Fox, R.C.: 1980, 'The Evolution of Medical Uncertainty', *Milbank Memorial Fund Quarterly*, 58 (1), 1–49.
6. Isambert, F.A.: 1983, 'Aux sources de la bioéthique', *Le Débat*, n°25 (mai), 83–99.
7. Isambert, F.A.: 1984, 'Quelques réflexions sur l'éthique dans le domaine biomédical', *Sciences Sociales et Santé*, 2 (3–4), 191–207.

8. Isambert, F.A.: 1986, 'Révolution biologique ou réveil éthique?', Cahiers S.T.S. (Science-Technologie-Société) n°11: *Ethique et Biologie*, Editions du CNRS, Paris, pp.9–41.
9. Jalbert, P. *et al.*: 1989, 'Genetic Aspects of Artificial Insemination with Donor Semen: The French CECOS Federation Guidelines', *American Journal of Medical Genetics*, 33, 269–275.
10. Moreno, J.D.: 1994, 'Consensus by Committee: Philosophical and Social Aspects of Ethics Committees', in this volume, pp. 145–162.
11. Novaes, S.: 1994 *Les Passeurs de Gamètes*, Nancy, Presses Universitaires de Nancy.
12. Novaes, S.: 1991, 'Le mouvement bioéthique et l'évolution des rapports de soins', Actes du Colloque *Le Retour de la Sociologie Morale: Autour des Travaux de François-André Isambert* (in press).
13. Novaes, S.: 1992, 'Etique et débat public: de la responsabilité médicale en matière de procréation assistée', Raisons Pratique, vol 3: *Pouvoir et légitimité: figures de l'espace public*, Paris, Editions de l'EHESS, pp. 155–176.
14. Selva, J. *et al.*: 1986, 'Genetic Screening for Artificial Insemination by Donor (AID): Results of a study on 676 semen donors', *Clinical Genetics*, 29, 389–396.
15. Wertz, D.C. and Fletcher, J.C. (eds.): 1989, *Ethics and Human Genetics: A Cross-Cultural Perspective*, Springer, Berlin, Heidelberg, New York.

ANDREAS VOẞ

"... AND THAT IS WHY I WOULD LIKE AS FEW PEOPLE TO BE INVOLVED AS POSSIBLE"

*Observations on the Possibilities Offered by Consensus Achievement Within the Field of the Human Reproductive Technologies**

I. INTRODUCTION

A sociologist writing on the topic "consensus" is influenced by images of man molded by contractualism. Contractualism takes for granted the human ability to behave in a reasonable and intersubjective manner, an ability bound to good sense. This behavior requires in turn a social setting consisting of a sensible oral or written exchange of arguments, and is thus tied to employment of the language.

Working through the various areas of human behavior, one is seldom confronted by the ability to exchange (and the necessity of exchanging) arguments sensibly; one is not even necessarily confronted by the employment of language. Human beings do not construct reality primarily through an oral or written exchange of arguments [3], but through intersubjective – often enough non-verbal – action. Their actions are not marked by the abstract category "good sense", but by the daily pressure to act. Good sense or being sensible are, at best, categories which may be present in the legitimation of past or future actions, but not in the daily actions taking place at the time in question.

> The nature of the protagonist, especially his temporality, and the nature of the reality surrounding him, especially its temporality, impose non-exceedable limits on the knowledge of the protagonist. The idea of action being unlimitedly sensible remains exactly that: an idea ([9], p. 90).

This is, of course, just as true for those acting within the field of human reproductive technology. Even if actions within the medical sphere involve styles of action which are far nearer to the style of institutional actions than to that of everyday actions, this still does not render

Kurt Bayertz (ed.), The Concept of Moral Consensus, 223–234.

the field of reproductive medicine automatically a field which is suited *per se* to social settings which enable a sensible, contractually oriented achievement of (moral) consensus. It is not of primary importance here whether the consensus "emerges from group interaction or is merely a coincidence of opinion" ([8], p. 161), nor whether smaller or larger groups are to be involved in the reaching of consensus, even if these differences are of principal sociological importance. In the next few pages I would like to put forward the following thesis: in the attempt to achieve consensus regarding the social acceptibility of the application of human reproductive technology, the involvement in the process of consensus achievement of those affected by the technology, based on the ethics of discourse [6] and supported by the principles of autonomy and self-determination ([2], p. 52), will meet with difficulties. The majority of unwantingly childless couples who take advantage of the possibilities offered by a course of medical treatment incorporating the reproductive technologies are not in a position to take part in the construction of a moral consensus concerning the application of reproductive technologies and the allied problems. This is not because the patients in a fertility clinic lack the will to exchange arguments sensibly, or even that they lack the intelligence necessary for an 'orderly debate', but because couples affected by unwanted childlessness do not use speech but silence to express their problem. These couples do not use the social structure of sensible debate to express symbolically their infertility; rather, in order to do this, they require the social settings which are bound to the discretion and isolation of medical institutions. A process of consensus achievement which, besides the participation of specialists (psychologists, philosophers, sociologists, theologans, lawyers, etc.), favors the participation of affected subjects within the framework of its debating circles, will in the case of reproductive medicine thus be destined to fail structurally.

In the following I shall present some examples of the social functions behind the medical treatment of unwanted childlessness, and, using these as a basis, shall define the superior social function of reproductive medicine. At the end of my paper I shall bring together these superior social functions of reproductive medicine and the social demands to which an achievement of consensus necessarily gives rise.

II. SOCIAL FUNCTIONS BEHIND THE MEDICAL TREATMENT OF UNWANTED CHILDLESSNESS

It is clear, from narrative interviews which we have carried out with unwantingly childless couples within the framework of a research project[1], that the vast majority of patients withold from their family, friends and acquaintances firstly the fact that they have come up against obstacles in the carrying out of their desire to have children, and, secondly, the fact that for this reason they are undergoing medical treatment. A 34 year-old woman who has been undergoing medical treatment (AID = Artificial Insemination from Donor) for six years relates, for example, the following:

> ... Nobody even knows that we are undergoing gynaecological treatment. ... I think it's terribly important that the child never finds out that he or she, er, that the father is not the biological father, or that others are the biological parents at all. I really would be scared of the child finding that out, *and that is why I would like as few people to be involved as possible....*

Yet it is not only the couples who try to conceive a child with the help of 'Artificial Insemination from Donor' who would like to restrict the circle of those in the know. Couples using homologous techniques or who are 'merely' undergoing hormone therapy prefer to remain silent about their problem and the fact that they are undergoing medical treatment, or at least insofar as they can manage it. In cases where the woman works, the employer often has to be informed, which the patients consider extremely unpleasant. A 39 year-old former AIH (AIH = Artificial Insemination from Husband) patient on the subject:

> ... Take Saturday, for instance, when the right time (for insemination) falls on a Saturday, and of course I have to work Saturdays. And then my boss said: what, now you have to go to the doctor on Saturdays too? And then I found it awful having to find a way to explain it to him on top of everything...

The person affected obviously does not feel happy talking the problem over with outsiders. Neither the family, nor friends or acquaintances, and certainly not the public are in a position to be an appropriate forum within which the unwantingly childless couple could express their problem. (Exceptions here too prove the rule [11].) Yet if both the private sphere and the public sphere are, to a certain extent, zones of silence for the unwantingly childless couple, then the institutional sphere is the only place left for them to discuss and overcome their problem. The majority of unwantingly childless couples in our society seek medical institutions to help them find a solution to their problem.

Using the expression 'the desire to have children', unwanted childlessness is very quickly categorized by those affected as a medical condition, and thus in need of medical attention. In other words, relatively early on unwantingly childless couples begin to see an illness as a way of describing and solving their problem. Couples in the early stages of treatment tend not to talk openly about having an illness, simply attempting to cover it up. In the main they present a picture which regards infertility as a 'minor disturbance' which may be done away with by taking medicine. Detailed analysis shows, however, that the infringements on the well-being of the affected subjects, to which they themselves have referred, can, even at the beginning of the treatment and especially in the women, already be so tremendous (cysts, skin irritations, OHS syndrome, etc.) that the picture which the patients themselves have of their inability to have children ('minor disturbance') cannot be considered objectively valid. A key term in this context is 'the desire to have children'. Patients use this term in such a way that it is syntactically and semantically directly aimed at an idea of illness. In other words, those affected use 'the desire to have children' as if it was the name of an illness. After naming this 'illness' the patients usually go on to tell of the events which were involved in the decision to start undergoing medical treatment. A 29 year-old woman who has been having treatment (IVF/ET) for five years reports the following:

> It's very simple... in 1985 we had the desire to have children, but it was much earlier that we spoke about the fact that we would like to start a family, and in 1985 we said, okay, let's be more concrete about things and perhaps see to it that we will be able to have children, and we, well, kind of sought, er, as it were, er, doctors, first of all our GP, and asked why it could be that there seems to be a delay...

A 37 year-old woman who has been having treatment (AID) for the past three years expresses the same as follows:

> ... Yeah, and then, I can't remember exactly when though, I'd have to work it out, thirty-two... maybe at the age of thirty-four we had the desire to have children. So first of all I went off to see the doctor, as one does...

Entering the realms of the medical sphere does not only offer the unwantingly childless couple the chance to have medically treated what they have experienced as 'delay' in the fulfillment of their desire to have children. The medical sphere additionally offers them a forum in which they no longer have to remain silent about their problem. They are able to express and present their problem verbally and non-verbally to the doctors, to the rest of the medical staff, and to the fellow patients in the waiting-room. Within the medical sphere the symbolic expression of the problem of unwanted childlessness is thus able to unfold to its full extent. This is a social function of considerable importance behind the medical treatment of unwanted childlessness.

Even though the 'healing' striven after (= fulfillment of the desire to have children) plays the most important role for the patients, the medical treatment of unwanted childlessness is additionally endowed with an abundance of non-medical social functions for them. A 36 year-old man who has been undergoing treatment (AIH) with his wife for three years tells, for example, how he interprets the long treatment as a personal, fateful test of his desire to have children.

> Sometimes I think of the fact that we've been waiting for three years now as a kind of test which there's no way of avoiding, regardless of whether one ought to have children, or not.

A woman who has been undergoing treatment (AID) for five years also refers to a significant pattern of fate ('God willing') in connection with her medical treatment.

> ... and so now we've got the chance, he (the husband is meant) will, God willing, be around during the pregnancy, God willing, and, er, God willing, perhaps for the birth too, if the birth goes normally, God willing, and then, er, I'll have a child which is at least fifty percent own flesh and blood...

Obviously we are dealing here with patterns of fate with one essential advantage for those affected. By turning to the medical sphere with their unfulfilled desire to have children, the affected subjects are saved from having to accept their fate as irrefutable, in isolation and in silence. Instead, they can actively try to influence their fate in social activity with others (medics, medical staff and fellow patients). A further social function behind the medical treatment of unwanted childlessness consists in removing from patients the burden of everyday activities which could lead to the solution of their problem. It is known, for example, that more than a few couples undergoing medical treatment have no or little sexual contact with each other, indeed sometimes doing away with it altogether. During our field studies in a West German family planning clinic we met one married couple, for example, who had told the doctor that they only rarely had sexual contact with one another. Both partners, who were organically healthy, demanded vehemently that the doctor undertake a homological insemination with the woman. The doctor countered this notion with the succinct sentence: *"Why don't you give it another go with sexual intercourse?"* yet continued to treat the couple. A problem which is very obviously not medical is thus, in this case, to be solved with medical treatment.

The social functions which human reproductive technology can assume within the dynamics of married relationships become even more obvious in heterological medical procedures. Heterological medical procedures offer couples, where one of the two is infertile, the chance to have a baby without having to be temporarily unfaithful or to risk divorce. It may be noted here that in former times infertility was socially and legally recognized, and by no means uncommon, grounds for divorce ([5], pp. 183ff.). The social function of medical treatment is here in particular that of relieving couples from the selection of sperm donors. The affected patients leave this selection entirely up to the medics. The patients usually demand that the whole process remain anonymous. A 29 year-old woman who has been undergoing treatment (AID) for three years, asked about the process of selecting a sperm donor, replies:

> I didn't have anything to do with that. We sent the people here a photo of my husband, and I presume the donor was selected accordingly.

A 36 year-old man, who has been undergoing treatment with his wife for two years, on the same topic of selecting a sperm donor:

> ... I don't know what my wife feels about it. Neither do I have a burning desire to probe deeper and find out exactly what has happened...

A 31 year-old man, who has been undergoing treatment (AID) with his wife for two years, refers to the topic of selecting a sperm donor as follows:

> Well, maybe to do with the person, the type of person I am, like the colour of my hair, er, maybe the age too, but I wouldn't like to know any more about him (the sperm donor is meant). ... In the end I'm just glad that there are people around who are prepared to do something like that, and that's why I personally don't really want to probe very deeply as to who he is, or to get a detailed description of him. It really doesn't interest me.

There can be no doubt that the anonymity of the sperm donor not only protects the donor himself (e.g., in the legal sense, regarding paternity suits) or the treating medic; it also, and especially, protects the relationship between the unwantingly childless couple. The act of matrimonial unfaithfulness can be avoided and, instead of going through the sexual act with a partner outside the marriage, a heterological insemination is carried out in the sterile atmosphere of a clinic – an atmosphere which systematically excludes eroticism. Medics themselves often use the neutrally sounding term "*Spendersamenbehandlung*" ("*donated sperm treatment*") when talking to patients about AID, thus making their own contribution to the all-round attempt to reduce emotions to a minimum.

The human reproductive technologies thus offer unwantingly childless couples a range of non-medical, social functions which render the seeking of the medical sphere attractive. The examples of such social functions which have been mentioned here may be summarized as follows:

1. First of all, the institutional medical sphere principally offers a forum of presentation for unwanted childlessness. It provides a forum for discussion of a problem which, as much in the private as in the public sphere, is not talked about by those affected.
2. Choosing an illness (key expression: "desire to have children") to represent the unwanted childlessness sees to it that infertility, which is categorized as fate-determined, may be overcome, not in an isolated manner but embedded in social actions, by those affected. The fate of the unfulfilled desire to have children is thus not experienced as something final but remains influenceable, at least within the framework of the medically dominated routines of interpretation and action.
3. The incorporation of human reproductive technology offers unwantingly childless couples the possibility to carry out activities, which would otherwise be bound to sexuality, within the practically 'sexually free bounds' of the clinic, and, to a certain extent, through a third party in the form of the medic. This is of particular advantage from the perspective of those affected when heterological methods are involved.

What superior social function of reproductive medicine may then be concluded from what has been said so far? The answer is obvious: choosing an illness and seeking the medical sphere, in order to symbolically express unwanted childlessness, enables those affected by it to break through the isolation in which they find themselves. Medical institutions incorporate a protected sphere, strictly separate from everyday life and with firmly established routines of interpretation and action, which enables the couples to put forward their problem in social interaction with others. This social function of reproductive medicine is quite independent of whether the treatment leads to the birth of a child or not. The seeking of the medical sphere by unwantingly childless couples thus sociologically has the important social function, far beyond all the 'rational' reasons given – primarily based on medical 'necessity' -, of enabling a problem which can neither privately nor publicly be adequately expressed to be put forward within a protected, institutional field. It is the medical institutions in our society which are, as shown by the choice taken by those affected, currently prepared to concern themselves to the required degree of intensity with unwantingly childless couples, and which are in a position to do so. It should be briefly noted at this point that, in an historical context, there is nothing structurally new in the active procedure of seeking a sphere outside everyday life when the problem at hand is unwanted childlessness. In nearly all cultural circles infertility has always been, and still is today, a 'reproach' which often requires the efforts of transcendental powers for its removal. Rachel, for example, turned to God because of her continuing unwanted childlessness: "And God remembered Rachel, and God hearkened to her, and opened her womb. And she conceived, and bare a son; and said, God hath taken away my reproach" (Moses 1:30; 22–24). There are countless historical instances of the application of non-everyday, magical practices in order to bring an end to infertility. Christian pilgrimages ([7], pp. 288ff.) were undertaken in order to achieve fertility, for example, and methods which derived from non-Christian oriented popular superstitions were similarly used ([1], pp. 1374ff.).

It is not my intention with this historical evidence to make a cheap and general equation between the acts of magical or godly powers and the acts of medics, the so-called 'demigods in white'. Rather, I would like here merely to point out that a calling on the gods or on magical powers and a visit to the fertility clinic do have at least one thing structurally in common: the problem of unwanted childlessness is removed from its everyday context and placed in the hands of a transcendental or an institutional sphere.

If up until now I have shown that the medical sphere is the place today where unwanted childlessness can find adequate expression through affected couples, then this does not mean that the 'best of all worlds' has thus been found. The possibility of expressing a problem within the isolated medical sphere naturally does not remove the suffering which is caused by the necessity of keeping silent in everyday life. On this subject, a 34 year-old woman who has been undergoing treatment (AID) for seven years:

Woman: "... I'm at the end of my tether (sobs). I've already told my husband that I don't want to go on with it all (sobs). I can't take a break either. I'm thirty-four already (sobs)."

Interviewer: "... What's the main problem?"

Woman: "I don't know... having to keep it to yourself, not being allowed to talk about it (sobs), having to be careful at home that nothing slips out. Only my sister and I know (blows her nose and sobs), my husband didn't want anybody to know, but I just couldn't take it any longer (sobs). I'm so disappointed that it doesn't work (sobs)... "

There is also a high price, both in health and socially, to be paid by the affected subjects for the seeking of the medical sphere and the expression of unwanted childlessness in routines of interpretation and action. One only has to think of the side effects of intensive hormone treatment or the consequences arising for the unwantingly childless couple from years of sexual intercourse at set times and on doctor's orders.

III. CONCLUSION

There is no doubt that affected subjects are able to live without a moral consensus concerning the application of human reproductive technology ([4], p. 23), but those currently in action always can, or rather have to. There is no room for moral reasoning when actions are currently taking place, due to the inherent pressure to adapt and to act ([10], pp. 10ff.).

Neither can there be doubt that the possibilities offered by the human reproductive technologies touch upon superindividual, social interests. What used to be taken for granted is turned upside down; one only has to think of the legal consequences of dividing genetic and child-bearing maternity in the case of an egg or embryo donation (who is legally the mother?), of the ethical problems connected with isolated embryos (when does human life begin?) or of the problems surrounding the possibilities of the cryopreservation of genetic matter (what will the consequences be for future generations?). Agreement of social relevance must be reached about such and further questions. A single consensus between the medics and their patients as argued by H. T. Engelhardt ([4], p. 34) is in no way sufficient. A socially relevant process of consensus achievement cannot be satisfied with consensus between the physicians and patients directly involved, but rather requires the cooperation of discussants who are neither directly nor indirectly involved in the medical treatment. At first glance, the ethics of discourse seems to offer a solution to the problem. This discipline requires a 'broad' involvement in the negotiating of the validity of norms, and demands that everybody, including those who could potentially be affected, participate:

> According to the ethics of discourse, a norm only has a claim to validity when all the people who could be affected by it reach (or would reach) agreement that the norm is valid, as participants in a practical debate. ... A 'real' argumentation is needed, in which those affected participate cooperatively. Only an intersubjective reasoning process can lead to an agreement of a reflexive nature: only then can the participants be sure that they have convinced themselves about something together. ([6], pp. 76f.)

What basic chances a concept such as this has for the negotiation of the validity of norms cannot be discussed here. It is possible to establish, however, that such a broad cooperation of all those potentially affected by reproductive medicine is a crass contradiction of the principle of discretion and the structural isolation of the medical sphere, both of which are of such great importance for the unwantingly childless couple. Anyone who incorporates unwantingly childless couples in the social structure of rational discourse because of the principles of autonomy, self determination and the resulting participation of all those affected in the process of achieving consensus, anyone who ultimately demands that they reason with others about the universalisability of their actions, in so doing destroys the single 'functioning' field of protection and expression which is currently available to unwantingly childless couples. Such consensus-achieving processes would undermine the social functions of reproductive medicine and thus ultimately drive the unwantingly childless couples into total isolation. The dilemma connected with the application of methods from the ethics of discourse in the case of reproductive medicine is that, although enough potentially affected subjects could be found to 'participate cooperatively in real argumentation', very few of those subjects actually affected would be able to participate in the 'intersubjective process of communication' because of social pressures and the allied silent behavior.

If, on the one hand, one is not prepared to let the medical sphere, separated from the rest of society and equipped with a kind of full power of attorney, define its own values and norms without external control, and if, on the other hand, a broad debate between all those actually and potentially affected is not possible, then other methods, which have not yet been discussed, have to be employed: one path which appears traversible, and which does not cut the ground from under the feet of those actually affected, as far as the social expression of unwanted childlessness is concerned, would be to guarantee the medical sphere's independence. In other words, those people acting within the medical sphere would not be burdened with the processes of achieving consensus, but for the application of human reproductive technologies a binding framework for medics and patients would be defined externally via political, democratic methods, a framework which appears to be socially responsible and which in addition may be politically attainable and administratively controllable. Following this concept, the fields of work of those representing the various sciences which should be present during democratic decision making, to advise as impartial objective factors, would come to the fore within this democratic process, remaining removed from the concrete and practical application of reproductive medicine.

However, the democratic process suggested here for the reaching of agreements no longer deals with 'consensus', as the papers by K. Bayertz [2], H.T. Engelhardt, Jr. [4] and J.D. Moreno [8] confirm.

Regardless of what kind of agreement is sought after in the case of reproductive medicine, one should not overestimate the possibilities of abstract, philosophically founded reasoning. It is not a dimension which is significantly existent in everyday human lives. At most there are still the possibilities of 'practical reasoning', and these are only definable "when an action (of described universal structure) is measured against what a concrete protagonist, a living human being, can do in a concrete, historical world, and how he can do it" ([9], p. 90).

Department of Sociology
FernUniversität Hagen
Germany

NOTES

* Translated into English by Sarah L. Kirkby, B.A. Hons. Exon.

[1] This research project is part of a 3-year study of the social conditions and the social consequences of reproductive medicine, sponsored by the Northrhine-Westphalian Ministry of Science. The project is being led by Prof. Dr. Hans-Georg Soeffner from the Open University of Hagen.

BIBLIOGRAPHY

1. Bächthold-Stäubli, H. (ed.): 1931/32, *Handwörterbuch des deutschen Aberglaubens*, de Gruyter, Berlin, Leipzig.
2. Bayertz, K.: 1994, 'The Concept of Moral Consensus. Philosophical Reflections', in this volume, pp. 41–57.
3. Berger, P.L.; Luckmann, T.: 1969, *Die gesellschaftliche Konstruktion der Wirklichkeit – Eine Theorie der Wissenssoziologie*, Fischer, Frankfurt/M.
4. Engelhardt, H.T., Jr.: 1994, 'Consensus: How Much Can We Hope for?', in this volume, pp. 19–40.
5. Fischer-Homberger, E.: 1988, *Medizin vor Gericht – Zur Sozialgeschichte der Gerichtsmedizin*, Luchterhand, Darmstadt.
6. Habermas, J.: 1983, 'Diskursethik – Notizen zu einem Begründungsprogramm', in J. Habermas, *Moralbewußtsein und kommunikatives Handeln,* Suhrkamp, Frankfurt/M., pp. 53–125.
7. Kriss, R.: 1958, *Wallfahrtsorte Europas*, Hornung, München.
8. Moreno, J.D.: 1994, 'Consensus by Committee: Philosophical and Social Aspects of Ethics Committees', in this volume, pp. 145–162.
9. Schütz, A., Luckmann, T.: 1984, *Strukturen der Lebenswelt*, Band 2, Suhrkamp, Frankfurt/M.
10. Soeffner, H.G.: 1989, *Auslegung des Alltags – Der Alltag der Auslegung – Zur wissenssoziologischen Konzeption einer sozialwissenschaftlichen Hermeneutik*, Suhrkamp, Frankfurt/M.
11. Sonnemann, S.: 1987, *Mein Kind ist ein Retortenbaby*, Rowohlt, Reinbek.

H. TRISTRAM ENGELHARDT, JR.

A SKEPTICAL POSTSCRIPT: SOME CONCLUDING REFLECTIONS ON CONSENSUS

From the foregoing essays, it should be clear that there is little or no substantive consensus regarding consensus. Nor is there about its implications. From everyday life, it should be clear as well that there is no consensus about the moral significance of sexuality or of third-party technologically-assisted reproduction. Different religious and cultural groups offer different understandings of why one ought to have children and how one may go about producing them. Yet, in nearly all of the essays, there is nearly a consensus that the agreement of individuals is important in framing morally authoritative public policy. Against the background of the foregoing essays, one can see why this should be so. If one cannot draw authority for common action from the will of God or from a content-full understanding of moral rationality, one can straightforwardly derive it from common agreement. The difficulty is then that one seems to have so little about which one in fact agrees. It is not just that there is substantive disagreement in large-scale secular societies regarding morality or the good life. There is also substantive disagreement about what is just, about how one should balance interests in equality and liberty. These latter disagreements seem to threaten the very possibility not only of a substantive, but even of a procedural ethic. It does not seem possible, not simply as a societal fact, but as a matter of philosophical principle to agree how one would go about agreeing. What is at stake is the possibility of a universal narrative, a morality of moral strangers, a morality that can be shared by individuals of different moral communities, as well as those of ill-defined, cosmopolitan moral inclinations. Within particular moral communities, there are substantive views regarding the probity of third-party technologically-assisted reproduction. The question is whether enough is shared so that moral strangers can still be bound together in a morality, although they have

Kurt Bayertz (ed.), The Concept of Moral Consensus, 235–240.

different substantive understandings of the good life, of morality, of what is just, and of what are appropriate ways of reproducing.

The special focus of this volume, namely, technologically-assisted reproduction, provides a test of the possibility and meaning of consensus in secular society. This cluster of case examples also invites an exploration of what it means to have a secular morality. The test and the exploration have implications that are broad and profound. If one cannot in the circumstances of large-scale secular states draw moral authority from the content-full voice of reason, the question of nihilism is unavoidable. The very possibility of a secular morality is brought into question. Here, one encounters the skeptical worry that there may be no secular ethics for applied ethics to deploy, that there is no general secular morality to give content and direction to a secular bioethics.

Yet, there are institutions that bind moral strangers simply in terms of the agreement of those who participate. The appeal to common consent to derive secular moral authority does not require consensus for its justification. It provides the basis for understanding the significance of the practices that bind moral strangers such as the practices of free and informed consent, the free market, and limited democracies. One need not have a general agreement about what one ought to do in medicine to act together with common authority. It is enough that particular physicians and patients agree to act together. No substantive understanding of what is appropriate or allowable is required. Authority for common action can be derived from the bare decision to collaborate. Each party may be moved to participate by motives the other does not affirm or perhaps even rejects. So, too, in the market, goods and services can trade without any content-full *a priori* notion of a fair or appropriate price. It is enough that the buyers and sellers agree to exchange. A fair price is the price of a sale in the absence of fraud, deceit, or overreaching. Buyers and sellers can each have quite different understandings of the significance of the goods and services traded. The market endorses only the price of the exchange.

There is also the possibility of limited democracy, one that precludes the use of unconsenting innocents, allows the enforcement of recorded contracts, and can justify, by whatever rules have been set in place, the deployment of common resources for common endeavors, including the creation of welfare rights. Those who work together can give an account of the authority of what they do together in terms of common agreement. Those who use others without their consent cannot in general secular terms show why they should not be visited with punitive or defensive force. They become outlaws from the community of peaceable moral strangers, from the morality that can bind moral strangers in a commonly justifiable fashion. Those who do not wish to participate in a project, and do not use others without their consent, can retreat and act with consenting others. The possibility of collaborating in terms of a generally defensible secular moral authority remains if there are robust rights to privacy, areas where individuals can act with consenting others, absent the interference of third parties, including the state. Because of the limited capacity to justify a content-full, common secular morality, rights to privacy will be salient, resources will be both societal and private, and the free choice of individuals will have central place. Such will be the case, not because individuals are valued, private property useful, or freedom more important than equality. It is rather that such is the structure that can be justified when the authority for common action comes from neither God nor reason.

Post-modernity, the collapse of the modern moral philosophical project to discover a canonical, content-full narrative or account of morality or justice, does not leave us without a thin structure binding moral strangers. The neutral language within which moral strangers can collaborate is not dependent on particular moral visions, not even the moral visions of Americans or Texians. The language or discourse that can bind moral strangers in ways that should be justified to such strangers is a moral or logical possibility. It does not depend on particular socio-cultural conditions. Its realization in this world is likely to be the final common pathway of disparate socio-historical and political conditions.

Within both communities of moral friends and societies of moral strangers, there are questions of morality and politics. It is just that within a community of moral friends (which may compass millions of like-minded individuals across the globe) politics is driven by a communality of values and understandings of authority such as those Aristotle envisioned for his polis (*Politics* 7.4.1326b; *Nicomachean Ethics* 9.10.1170b). The level of politics that binds moral strangers with secular moral authority can presuppose at most shared procedures for resolving moral controversies and at least a grammar for a common moral language. Any particular content would at once be alien to many and without binding moral force.

Also, vague, general, or unstructured transcendental longings or yearnings should not be confused with a consensus regarding religious roots or religious direction, if by the latter one means a definitive moral content or context such as one finds within intact religious communities, where one can identify who is a moral authority, as well as who is in moral authority. No one, or hardly anyone, lives outside the guidance of some moral context, however sparse, fragmentary, and incoherent. Still, an overlapping of fragments of moral visions is not the same as common agreement or consensus. However, in politics various minority views can be built into a political collage or bricollage of power that can effectively take control of political power, but this is not a consensus.

In medicine in general, and in reproductive medicine in particular, where moral divergence is often deep and significant, it is important both to understand where consensus is possible and the limits of its moral significance. If societies create publicly funded reproductive interventions which some moral communities find morally opprobrious (which will in almost all circumstances be the case), there must be the possibility for those who are members of such communities both to refuse such care as well as to condemn it openly as immoral. It may often be possible to create a sufficient consensus or rather a sufficient political concurrence so as to authorize the funding from common resources of endeavors that many will find improper. Such will be justifiable only if a line is sought between public and private resources, and if space is allowed to create from private resources alternative approaches.

The impoverished moral authority of secular, limited democracies will not be celebrated by those who wish to use the state to impose either religious mores or a particular content-full secular mores or ideology. It will not be possible on general secular grounds to forbid access to abortion, though many believers recognize abortion to be a serious moral evil (as does the author of this post-script). Nor will it be possible for secular protagonists with particular moral views to forbid women from acting as surrogate mothers for hire (a practice this author also condemns on religious grounds). Nor, importantly, will it be possible to forbid peaceable moral denunciations of such practices. A peaceable post-modernity must constrain coercive acts. However, it may not proscribe peaceable, public moral condemnations. The toleration that must constitute the foundational fabric of secular societies compassing communities of diverse moral vision should support peaceable negotiation without seeking to undermine the content-full commitments that give character and substance to the moral life. The peaceable toleration that ought to characterize a secular society compassing communities of divergent viewpoints should restrain direct, coercive acts of force, but it must tolerate mutual condemnations and strident moral denunciations. It must also protect against an overriding consensus becoming coercive by recognizing rights to privacy even in the face of substantial contrary majorities.

The recognition that one may not with general secular authority coercively constrain others to accept the content-full commitment of one's own moral community does not require abandoning one's own commitments in favor of a neutral vacuous moral language. Indeed, one may by witnessing to one's convictions attempt to convert those who have rejected grace while at the same time collaborating with them. Indeed, the reader should be warned that the author holds there is much more to the good life and to a proper ethos of reproduction than can be put in general, secular terms. The author is, after all, a believing Christian with Orthodox understandings of the moral constraints set on reproductive interventions. What is offered is not a proposal about what is good to do, but a reconstruction of what can be justified in general secular terms when there is in fact no agreement regarding the good or the just. This public confession is offered not simply to prevent misunderstandings regarding the views endorsed. It is also a warning about what it is to come to terms with bioethics and health care policy in large-scale secular states. There will be clashing moralities and communities to encompass that do not share a substantive consensus. Yet there will also be the possibility to cooperate peaceably with common moral authority both in mutual protection, as well as in the creation of limited endeavors of common solidarity, as, for example, in the creation of refusable welfare rights.

This account is not advanced as a proposal I celebrate. It is offered as a reconstruction of what we find when numerous particular moralities meet and reason does not disclose outside of any particular moral vision how one ought content-fully to understand the good life, morality, or justice. There are crucial differences at stake here among (1) the moralities of traditional and religious communities, (2) the dominant, secular, liberal ideology of any particular secular society, and (3) the sparse moral framework within which moral strangers can communicate. If these differences are not observed and recognized, one will confuse the dominance of a particular secular ideology (its widespread acceptance, its control of the governmental apparatus, and its dominance of the media) with its moral authority. One will as well not recognize the similarities between liberal ideologies and religious orthodoxies. Both liberal ideologies and religious orthodoxies involve content-full moral visions. Each claims an authority that cannot be justified to moral strangers in general terms. The more one comes to recognize the particularity of all content-full liberal ideologies, as well as the content-less character of the moral language that can bind moral strangers, the more the secular moral authority of the state born will be brought into question.

The essays in this volume illustrate the diversity of moral visions, as well as the need to collaborate across diversity without denying its existence. The essays have shown as well the need to attend to the ambiguities of "consensus" and to find its proper roles in warranting common actions. If these ambiguities are not noted, the appeal to consensus is an appeal to an *ignis fatuus*, a phantasm, an understandably attractive and deceptive myth. There is in fact no moral consensus or moral vision shared by those who live within inevitably pluralistic large-scale states. The crisis of morality and secular political authority announced by post-modernity discloses the impossibility of discovering that by which a moral consensus could be justified as a matter of secular morality in large-scale states. Still, the belief in a consensus is a compelling self-deception. The appeal to consensus suggests a communality that would solve many theoretical and political problems. If it did exist, it would provide authoritative direction. Its absence is therefore mourned by many and defensively denied by even more. As an unqualified term, it remains, however, an instrument of demagogues, realpolitik, and power politics within pluralist democracies.

Consensus in the sense of political concurrence is integral to fashioning public direction for the use of public funds. Such a consensus is not one that can ground a content-full public morality to which all should subscribe. Here the best one can hope for is a commitment to a fabric of formal negotiation that recognizes a robust line between the private and public life, between private and public resources. Through the recognition of such limits, we not only will be able to act in public endeavors with an authority that should be justified to all, but should in addition be able to live in peace and in the integrity of our moral convictions. There are concepts of consensus that can guide and help; there are others that can misguide and harm. Among these families of notions of consensus, one must move with care so that one selects correctly in order to support the moral life, rather than to be enslaved to notions that obscure rather than enlighten.

NOTES ON CONTRIBUTORS

Kurt Bayertz, Dr. phil., is Professor of Philosophy at the University of Münster, Germany.

Alberto Bondolfi, Dr. theol., is a permanent staff member of the Institute for Social Ethics of the University of Zurich, President of the Swiss Society for Biomedical Ethics, and a member of the central ethics committee of the Swiss Academy of Medical Sciences, Switzerland.

Wolf-Michael Catenhusen is Member of the German Bundestag and Chairman of the Bundestag Committee on Research, Technology and Technology Assessment, Bonn, Germany.

James Childress, Ph.D., is Kyle Professor of Religious Studies and Professor of Medical Education, University of Virginia, Charlottesville, Virginia.

H. Tristram Engelhardt, Jr., Ph.D., M.D., is Professor, Department of Medicine, as well as Community Medicine and Obstetrics and Gynecology, Baylor College of Medicine; also Professor, Department of Philosophy, Rice University Adjunct Research Fellow, Institute of Religion and Member, Center for Ethics, Medicine and Public Issues, Houston, Texas.

Henk A.M.J. ten Have, M.D., Ph.D., is Professor of Medical Ethics, Department of Ethics, Philosophy and History of Medicine, Faculty of Medical Sciences, Catholic University of Nijmegen, Nijmegen, The Netherlands.

Kurt Bayertz (ed.), The Concept of Moral Consensus, 241–242.

Ludger Honnefelder, Dr. phil., is Professor and Director, Department of Philosophy, University of Bonn, Germany.

Helga Kuhse, Ph.D., is Director, Centre for Human Bioethics, Monash University, Clayton, Melbourne, Victoria, Australia.

Jonathan D. Moreno, Ph.D., is Professor of Pediatrics and of Medicine; and Director of the Division of Humanities in Medicine, State University of New York Health Science Center at Brooklyn, New York.

Simone Novaes, Doctorat de troisième cycle, does sociological research at the Centre de Sociologie de l'Ethique (EHESS-CNRS), Paris, France.

Larry Tancredi, M.D., J.D, is Professor of Clinical Psychiatry, and Director, Health Law Program, University of Texas, Health Science Center at Houston, Texas.

Peter Weingart is Professor, Department of Sociology, and Science Studies Unit, University of Bielefeld, and Director of Center for Interdisciplinary Research (ZiF), Bielefeld, Germany.

Carl Wellman, Ph.D., is Professor, Department of Philosophy, Washington University, St. Louis, Missouri.

Andreas Voß is Dr. rer. soz., Department of Sociology, FernUniversität Hagen, Germany.

INDEX

Philosophy and Medicine

1. H. Tristram Engelhardt, Jr. and S.F. Spicker (eds.): *Evaluation and Explanation in the Biomedical Sciences.* 1975 ISBN 90-277-0553-4
2. S.F. Spicker and H. Tristram Engelhardt, Jr. (eds.): *Philosophical Dimensions of the Neuro-Medical Sciences.* 1976 ISBN 90-277-0672-7
3. S.F. Spicker and H. Tristram Engelhardt, Jr. (eds.): *Philosophical Medical Ethics: Its Nature and Significance.* 1977 ISBN 90-277-0772-3
4. H. Tristram Engelhardt, Jr. and S.F. Spicker (eds.): *Mental Health: Philosophical Perspectives.* 1978 ISBN 90-277-0828-2
5. B.A. Brody and H. Tristram Engelhardt, Jr. (eds.): *Mental Illness.* Law and Public Policy. 1980 ISBN 90-277-1057-0
6. H. Tristram Engelhardt, Jr., S.F. Spicker and B. Towers (eds.): *Clinical Judgment: A Critical Appraisal.* 1979 ISBN 90-277-0952-1
7. S.F. Spicker (ed.): *Organism, Medicine, and Metaphysics.* Essays in Honor of Hans Jonas on His 75th Birthday. 1978 ISBN 90-277-0823-1
8. E.E. Shelp (ed.): *Justice and Health Care.* 1981 ISBN 90-277-1207-7; Pb 90-277-1251-4
9. S.F. Spicker, J.M. Healey, Jr. and H. Tristram Engelhardt, Jr. (eds.): *The Law-Medicine Relation: A Philosophical Exploration.* 1981 ISBN 90-277-1217-4
10. W.B. Bondeson, H. Tristram Engelhardt, Jr., S.F. Spicker and J.M. White, Jr. (eds.): *New Knowledge in the Biomedical Sciences.* Some Moral Implications of Its Acquisition, Possession, and Use. 1982 ISBN 90-277-1319-7
11. E.E. Shelp (ed.): *Beneficence and Health Care.* 1982 ISBN 90-277-1377-4
12. G.J. Agich (ed.): *Responsibility in Health Care.* 1982 ISBN 90-277-1417-7
13. W.B. Bondeson, H. Tristram Engelhardt, Jr., S.F. Spicker and D.H. Winship: *Abortion and the Status of the Fetus.* 2nd printing, 1984 ISBN 90-277-1493-2
14. E.E. Shelp (ed.): *The Clinical Encounter.* The Moral Fabric of the Patient-Physician Relationship. 1983 ISBN 90-277-1593-9
15. L. Kopelman and J.C. Moskop (eds.): *Ethics and Mental Retardation.* 1984 ISBN 90-277-1630-7
16. L. Nordenfelt and B.I.B. Lindahl (eds.): *Health, Disease, and Causal Explanations in Medicine.* 1984 ISBN 90-277-1660-9
17. E.E. Shelp (ed.): *Virtue and Medicine.* Explorations in the Character of Medicine. 1985 ISBN 90-277-1808-3
18. P. Carrick: *Medical Ethics in Antiquity.* Philosophical Perspectives on Abortion and Euthanasia. 1985 ISBN 90-277-1825-3; Pb 90-277-1915-2
19. J.C. Moskop and L. Kopelman (eds.): *Ethics and Critical Care Medicine.* 1985 ISBN 90-277-1820-2
20. E.E. Shelp (ed.): *Theology and Bioethics.* Exploring the Foundations and Frontiers. 1985 ISBN 90-277-1857-1
21. G.J. Agich and C.E. Begley (eds.): *The Price of Health.* 1986 ISBN 90-277-2285-4
22. E.E. Shelp (ed.): *Sexuality and Medicine.*
    Vol. I: Conceptual Roots. 1987 ISBN 90-277-2290-0; Pb 90-277-2386-9

Philosophy and Medicine

23. E.E. Shelp (ed.): *Sexuality and Medicine.*
Vol. II: Ethical Viewpoints in Transition. 1987
ISBN 1-55608-013-1; Pb 1-55608-016-6
24. R.C. McMillan, H. Tristram Engelhardt, Jr., and S.F. Spicker (eds.): *Euthanasia and the Newborn.* Conflicts Regarding Saving Lives. 1987
ISBN 90-277-2299-4; Pb 1-55608-039-5
25. S.F. Spicker, S.R. Ingman and I.R. Lawson (eds.): *Ethical Dimensions of Geriatric Care.* Value Conflicts for the 21th Century. 1987
ISBN 1-55608-027-1
26. L. Nordenfelt: *On the Nature of Health.* An Action- Theoretic Approach. 1987
ISBN 1-55608-032-8
27. S.F. Spicker, W.B. Bondeson and H. Tristram Engelhardt, Jr. (eds.): *The Contraceptive Ethos.* Reproductive Rights and Responsibilities. 1987
ISBN 1-55608-035-2
28. S.F. Spicker, I. Alon, A. de Vries and H. Tristram Engelhardt, Jr. (eds.): *The Use of Human Beings in Research.* With Special Reference to Clinical Trials. 1988 ISBN 1-55608-043-3
29. N.M.P. King, L.R. Churchill and A.W. Cross (eds.): *The Physician as Captain of the Ship.* A Critical Reappraisal. 1988 ISBN 1-55608-044-1
30. H.-M. Sass and R.U. Massey (eds.): *Health Care Systems.* Moral Conflicts in European and American Public Policy. 1988 ISBN 1-55608-045-X
31. R.M. Zaner (ed.): *Death: Beyond Whole-Brain Criteria.* 1988
ISBN 1-55608-053-0
32. B.A. Brody (ed.): *Moral Theory and Moral Judgments in Medical Ethics.* 1988
ISBN 1-55608-060-3
33. L.M. Kopelman and J.C. Moskop (eds.): *Children and Health Care.* Moral and Social Issues. 1989 ISBN 1-55608-078-6
34. E.D. Pellegrino, J.P. Langan and J. Collins Harvey (eds.): *Catholic Perspectives on Medical Morals.* Foundational Issues. 1989 ISBN 1-55608-083-2
35. B.A. Brody (ed.): *Suicide and Euthanasia.* Historical and Contemporary Themes. 1989 ISBN 0-7923-0106-4
36. H.A.M.J. ten Have, G.K. Kimsma and S.F. Spicker (eds.): *The Growth of Medical Knowledge.* 1990 ISBN 0-7923-0736-4
37. I. Löwy (ed.): *The Polish School of Philosophy of Medicine.* From Tytus Chałubiński (1820–1889) to Ludwik Fleck (1896–1961). 1990
ISBN 0-7923-0958-8
38. T.J. Bole III and W.B. Bondeson: *Rights to Health Care.* 1991
ISBN 0-7923-1137-X
39. M.A.G. Cutter and E.E. Shelp (eds.): *Competency.* A Study of Informal Competency Determinations in Primary Care. 1991 ISBN 0-7923-1304-6
40. J.L. Peset and D. Gracia (eds.): *The Ethics of Diagnosis.* 1992
ISBN 0-7923-1544-8

Philosophy and Medicine

41. K.W. Wildes, S.J., F. Abel, S.J. and J.C. Harvey (eds.): *Birth, Suffering, and Death.* Catholic Perspectives at the Edges of Life. 1992
ISBN 0-7923-1547-2; Pb 0-7923-2545-1
42. S.K. Toombs: *The Meaning of Illness.* A Phenomenological Account of the Different Perspectives of Physician and Patient. 1992
ISBN 0-7923-1570-7; Pb 0-7923-2443-9
43. D. Leder (ed.): *The Body in Medical Thought and Practice.* 1992
ISBN 0-7923-1657-6
44. C. Delkeskamp-Hayes and M.A.G. Cutter (eds.): *Science, Technology, and the Art of Medicine.* European-American Dialogues. 1993 ISBN 0-7923-1869-2
45. R. Baker, D. Porter and R. Porter (eds.): *The Codification of Medical Morality.* Historical and Philosophical Studies of the Formalization of Western Medical Morality in the Eighteenth and Nineteenth Centuries, Volume One: Medical Ethics and Etiquette in the Eighteenth Century. 1993 ISBN 0-7923-1921-4
46. K. Bayertz (ed.): *The Concept of Moral Consensus.* The Case of Technological Interventions in Human Reproduction. 1994 ISBN 0-7923-2615-6
47. L. Nordenfelt (ed.): *Concepts and Measurement of Quality of Life in Health Care.* 1994 ISBN 0-7923-2824-8

KLUWER ACADEMIC PUBLISHERS – DORDRECHT / BOSTON / LONDON

MIX
Papier aus verantwortungsvollen Quellen
Paper from responsible sources
FSC® C105338

If you have any concerns about our products,
you can contact us on
ProductSafety@springernature.com

In case Publisher is established outside the EU,
the EU authorized representative is:
Springer Nature Customer Service Center GmbH
Europaplatz 3, 69115 Heidelberg, Germany

Printed by Libri Plureos GmbH
in Hamburg, Germany